TOPOLOGY
for beginners

TOPOLOGY
for beginners

Noor Muhammad
Asghar Qadir
Imran Parvez Khan

OXFORD
UNIVERSITY PRESS

Oxford University Press is a department of the University of Oxford.
It furthers the University's objective of excellence in research, scholarship,
and education by publishing worldwide. Oxford is a registered trade mark of
Oxford University Press in the UK and in certain other countries

Published in Pakistan by
Oxford University Press
No. 38, Sector 15, Korangi Industrial Area,
PO Box 8214, Karachi-74900, Pakistan

© Oxford University Press 2022

The moral rights of the authors have been asserted

First Edition published in 2022

All rights reserved. No part of this publication may be reproduced, stored in
a retrieval system, or transmitted, in any form or by any means, without the
prior permission in writing of Oxford University Press, or as expressly permitted
by law, by licence, or under terms agreed with the appropriate reprographics
rights organization. Enquiries concerning reproduction outside the scope of the
above should be sent to the Rights Department, Oxford University Press, at the
address above

You must not circulate this work in any other form
and you must impose this same condition on any acquirer

ISBN 978-969-7341-82-5

Typeset in Computer Modern Roman (cmr)
Printed on 68gsm Offset Paper

Printed by Delta Dot Technologies (Pvt.) Ltd., Karachi

Illustrations by Imran Parvez Khan
Cover illustration: Einstein-Rosen Bridge

In the Memory of
Rabiya Asghar Qadir
Beloved wife of Asghar Qadir

Contents

Preface		**xi**
1	**Introduction**	**1**
	1.1 The Broad Context of Topology: Mathematics	2
	1.2 The Development of Mathematics	3
	1.3 The Advent of Topology	6
	1.4 The Mathematical Background for Topology	8
	1.5 A Psychological Aside	10
	1.6 Topological Constructions	11
	1.7 The Need to Use the Language of Sets	14
	1.8 A Diversion into Logic	15
	1.9 The Language of Sets	17
	1.10 Counting the Elements of a Set...	18
	1.11 Uncountable Sets and Transfinite Numbers	22
	1.12 More General Topological Transformations	25
	1.13 Some Additional Remarks About Sets	28
2	**Topological Spaces**	**30**
	2.1 The Definition of a Topological Space	30
	2.2 Examples of Topologies	32
	2.3 The Interior, Closure and Boundary...	35
	2.4 Neighborhoods and Neighborhood Systems	37
	2.5 Isolated Points	39
	2.6 Some Simple Topological Theorems	41
	2.7 Topology in Terms of Closed Sets	42
	2.8 Limit Points, the Derived Set and...	43
	2.9 Dense Sets and Separable Spaces	46
	2.10 Topological Bases	48
	2.11 Criteria for Topological Bases	49
	2.12 Local Bases	51
	2.13 Relative or Induced Topologies	52
	2.14 Exercises	55
3	**Metric Spaces**	**57**
	3.1 The Definition of a Metric	59
	3.2 The Topology Induced by a Metric	64
	3.3 Equivalent Topologies Generated by Metrics	69
	3.4 Formulation with Closed Sets	70

3.5	Complete Metric Spaces	72
3.6	Characterization of Completeness	74
3.7	Further Theorems Related to Completeness	79
3.8	Contraction Mapping Principle and Applications	82
3.9	The Completion of a Metric Space	85
3.10	Exercises	88

4 Continuous Functions and Homeomorphisms — 90

4.1	Continuous Functions on Topological Spaces	90
4.2	Some Theorems About Continuous Functions	93
4.3	Homeomorphisms	98
4.4	Open and Closed Continuous Functions	100
4.5	Topological Properties and Homeomorphisms	104
4.6	Exercises	106

5 The Separation Axioms — 108

5.1	The Three Separation Axioms	109
5.2	Theorems About Separability	112
5.3	Product Topologies and Product Spaces	114
5.4	Hausdorff Spaces	116
5.5	T_3-Spaces	120
5.6	Regular Spaces	121
5.7	Theorems About Regular Spaces	122
5.8	Normal Spaces	124
5.9	Exercises	128

6 Compact Spaces — 130

6.1	The Standard Definition of Compactness	131
6.2	Characterizations of Compact Spaces	132
6.3	Construction of Compact Subspaces	133
6.4	Theorems About Compact Spaces...	135
6.5	Compactness in Metric Spaces	136
6.6	Equivalence of Definitions of Compactness	140
6.7	Compactness and Completeness	142
6.8	Local Compactness	144
6.9	Exercises	146

7 Connectedness — 147

7.1	Connected Spaces	147
7.2	Further Theorems for Connected Spaces	149
7.3	Connected Subspaces	151
7.4	Other Characterizations	153
7.5	Connectedness of Families of Sets	155
7.6	Path Connectedness	157
7.7	The Components of a Space	160
7.8	Total Disconnectedness	163
7.9	Locally Connected Spaces	165
7.10	Exercises	169

8 Further Directions in Topology and Applications — 171
  8.1 Further Directions in Topology — 171
  8.2 Application of Multiply Connected Spaces to Wormholes — 179
  8.3 Application of Winding Numbers to Magnetic Fields and Monopoles — 181
  8.4 Application of Connected Sets to Digital Image Processing (DIP) — 183
  8.5 Application of Graphs to Electronic Chip Design — 184

Appendix — 187

Bibliography — 188

Index — 190

Preface

This book is meant to be a first introduction to the subject of Topology. It is based on a one semester course given separately by Noor Mohammad (NM) and Asghar Qadir (AQ) at the Mathematics Department, Quaid-i-Azam University, Islamabad, Pakistan. It was generally offered to the first semester M.Sc. class. The Pakistani M.Sc. of the time was about the same standard as the British B.Sc. Hons. or the American BS, and was a two year course. It has since been converted to the senior years of a four year BS. As such, the book should be appropriate as a first introduction to the subject at the undergraduate level in most universities. No special background is required to follow the book beyond the British 'O' level or American High School. However, a slightly greater degree of mathematical *maturity* is required.

It is not written like a usual textbook on Topology. Let us explain in what ways it is different, First, topologists have rightly learned to mistrust the use of "pictures", or diagrams, which may often be quite misleading. However, that approach tends to make the subject very abstract and hence abstruse. Early intuition is firmly based in geometrical concepts which are best explained by "pictures". Students need to be gradually "weaned away" from using them rather than plunged into a bewildering world of great abstraction. Second, we feel that definitions are not enough to *explain* concepts. As such we use a lot of examples for each concept introduced. Further, as explained in the Introduction in more detail, we believe that examples are inadequate and that they must be supplemented by *counterexamples* to clarify the concepts. Third, the *raison d'etre* for any definition really comes out by exploring changes in it. Where possible we have tried to do so. Fourth, the subject, which was regarded as one of the last bastions of the purest of pure mathematics, has begun to be applied. We indicate some of its applications here. Fifth, the style of presentation is *extremely* informal, not to say *downright chatty*. We, as serious mathematicians, would not have had it so, but that is the way the book insisted on being written and we had to go along with it. For this we must present our excuses.

The fact is that one of us (AQ) was very strongly influenced by John Archibald Wheeler, Roger Penrose and Isaac Asimov. John had demonstrated the importance of an attractive writing style in serious scientific and professional work. Roger had made dialogues acceptable in a serious scientific work — "The Emperor's New Mind"[Penrose, 1989]. Asimov had shown how *any subject* could be made absorbing, regardless of how dull it had been considered. The blame (if any) for the style adopted, must be laid squarely at their door. The credit, if any, the authors will most happily share.

Finally, we would like to acknowledge, apart from the three people mentioned above, our students on whom we tried out the book; our colleagues for their suggestions and advice; the Quaid-i-Azam University for the opportunity; King Fahd University of Petroleum and Minerals for providing financial support in the preparation of this book and our families for their understanding and endurance. The long line of mathematicians, on whose shoulders we stand to survey this field, we can not begin to acknowledge.

Noor Mohammad
Asghar Qadir
Islamabad 1997

Postscript

We must also apologize for leaving the book unpublished for so long. Part of the blame goes to NM for dying so inconsiderately, when the book just needed the final proof-reading touches and the other part on AQ for not getting on with the job and slacking without the excuse of dying.

This book started as a series of lectures delivered by one of us (AQ) at the Mathematics Department of the Quaid-i-Azam University, Islamabad in 1994, which an exceptional student named Amer Iqbal, attended. He went on to obtain his PhD thesis in Superstrings at MIT. He was subsequently recommended by the famous Harvard String theorist, Camrun Vafa, whose group he worked in, as being the mainstay of that group because of his extremely strong background in Topology. We are grateful to him for proof-reading the manuscript of this book and giving useful comments.

Asghar Qadir
Imran Parvez Khan
Islamabad 2022

List of Figures

1.1	(a) An example of Riemannian geometry. (b) An example of Lobachevskian geometry	6
1.2	The seven bridges problem	6
1.3	A topological representation of the seven bridges problem.	7
1.4	The four color problem criteria	7
1.5	The four color problem statement, with colors replaced by numbers here.	8
1.6	The component of a vector **a** along **b** is $a\cos\theta$.	10
1.7	(a) A rubber string and band showing the difference of ordering. (b) The connectedness of the string and band. (c) The difference between them on cutting	11
1.8	(a) Ordering of directions in rectangle and cylinder. (b) Connectedness of the rectangle and cylinder. (c) The difference between the two on cutting.	12
1.9	(a) Construction of Möbius strip. (b) A 'seeing is believing' proof that Möbius strip has only one side. (c) On cutting, the Möbius strip does not split in two, as expected of an ordinary strip.	13
1.10	A Klein Bottle	14
1.11	A Torus	14
1.12	Increasing the length and decreasing the breadth of a rectangle can leave the area fixed while increasing the perimeter indefinitely.	14
1.13	Koch curve (a snow flake).	15
1.14	Geometric procedure to count rationals	21
1.15	Any line segment consists of infinite many points.	22
1.16	Geometric construction to locate the irrational number $\sqrt(2)$ on the real line.	23
1.17	(a) Mapping from an arc to a straight line. (b) Mapping from a curve of arbitrary shape to another curve of arbitrary shape.	26
1.18	Stereographic projection of a circle on a line.	26
1.19	Geometry of stereographic projection	27
1.20	Venn diagrams capturing several relations plausible among two subsets of a set X.	29
2.1	The union of $A = (0,2)$ and $B = (1,3)$ is $C = (0,3)$ which is an open interval.	31
2.2	Open sets in the real line, showing that an infinite intersection of open sets can be closed.	32
2.3	The A–inclusion topology in Venn diagrams.	33
2.4	The A–exclusion topology in Venn diagrams.	34
2.5	Only one proper subset of $\mathbb{R}$ is called "open" here. (a) The "right ray topology". (b) The "left ray topology".	34
2.6	Using Venn diagrams to construct the boundary of a set.	36
2.7	Interior and boundary of a disc	37
2.8	Neighborhood of a point	38
2.9	Example of an Isolated point	39
2.10	Isolated points of $\mathbb{Z}$ in $(\mathbb{R}, \tau_u)$	40

3.1	Distance between functions	59
3.2	Metric defined by max function	61
3.3	The open sphere in $\mathbb{R}^3$	64
3.4	Open sphere for ρ_1	65
3.5	Open sphere for ρ_2	66
3.6	Open sphere in $C([0,1]$	66
3.7	Open set in metric space	67
3.8	Union of open spheres is open	68
3.9	Equivalence of open square and open disc	69
3.10	Closed sphere is closed set	72
3.11	Approximation of a function by a nowhere differentiable function	81
3.12	A continuous but nowhere differentiable function	82
3.13	Existence of a fixed point	83
3.14	Scaled fixed point	83
3.15	A fixed point may or may not exist if the interval is not unit	84
4.1	Two discontinuous functions	91
4.2	Continuity of the modulus function	95
4.3	Continuity of piecewise defined functions	97
4.4	The step function is discontinuous at $x = 0$.	98
4.5	Injective and Surjective mappings	98
4.6	(a) Graph of $\sin x$. (b) Graph of $\sin^{-1} x$.	100
4.7	Open mapping	101
4.8	Openness for composite functions, with a surjection	102
4.9	Openness for composite functions, with an injection	103
5.1	Metric spaces are T_2	110
5.2	Separation in open infinite strips	111
5.3	Separation in open squares	111
5.4	Open rectangles	115
5.5	Plane minus the first quadrant	116
5.6	A non-Hausdorff space	117
5.7	The T_3 condition	120
5.8	Distance from a point to a plane.	123
6.1	Some members of $\mathcal{O}_1$ and $\mathcal{O}_3$ are displayed. As expected, $\mathcal{O}_3 \subset \mathcal{O}_1$.	131
6.2	Compactness of $(0,1)$.	134
6.3	As $Y \subset X$, $x_0 \in X \smallsetminus Y$ will live in the annular region as shown.	135
6.4	Construction of a sequence of open spheres is displayed. The limit point of the set $A, x \in S_{1/k}$. Hence the subsequence $\{x_{n_k}, x_{n_{k+1}}, \ldots\}$ contained in $S_{1/k}$ converges to x.	137
6.5	Example of an ϵ-net	138
6.6	Sequence of sets of decreasing diameters.	141
7.1	Lines L and M passing through the origin are connected subspaces of $\mathbb{R}^2$.	151
7.2	Intersection of two connected sets.	155
7.3	A *path* between two points	157
7.4	Two points in $\mathbb{R}^n \smallsetminus \{O\}$.	158
7.5	Topologist's sine curve	159

7.6	(a) Equally spaced horizontal lines. (b) Equally spaced vertical lines.	160
7.7	Cantor set	164
7.8	Comb space	166
7.9	The digital line. Some of the open sets defined in Eq. (7.35) are shown.	168
8.1	Path between two points on a real line	172
8.2	Path between two points on a circle.	172
8.3	Path between two points on a plane.	172
8.4	Path between two points on surface of a sphere	173
8.5	Path between two points on surface of a cylinder	173
8.6	Path between two points on an annular disk.	174
8.7	$\mathbb{R}^2 \smallsetminus \{O\}$ is topologically equivalent to a cylinder.	174
8.8	A plane cut into two pieces.	175
8.9	A sphere cut into two pieces.	175
8.10	A cylinder cut into two pieces.	176
8.11	(a) A torus cut into two pieces vertically. (b) A torus cut into two pieces horizontally.	176
8.12	Paths between two points having different winding numbers	177
8.13	Lorenz attractor	178
8.14	Mandlebrot set	178
8.15	Effect of placing a ball in rubber-sheet geometry	179
8.16	Journey of a ball through the Einstein-Rosen bridge. (a) The ball has just broken through. (b) The narrowing of the throat as the ball gets denser.	180
8.17	(a) A worm eating through the apple. (b) Star observed through a wormhole	180
8.18	Journey of a neutral, a negatively charged, and a positively charged particle on a cylinder surrounding a bar magnet.	181
8.19	Working of a SQUID, as the magnet goes through loop in (a), and a trace is drawn on the drum, as shown in (b). The pattern of the trace corresponding to the two options is shown in (c).	183
8.20	$4-nbd$ and $8-nbd$ in a Digital space.	183
8.21	Example 4−connectedness and 8−connectedness.	184
8.22	Pixel reduction in DIP as one downsizes from the full square in (a) to the one containing only the boundary as shown in (b).	184
8.23	Comparison between a PCB shown in (a) and a graph shown in (b).	185
8.24	Example of a bipartite graph.	185
8.25	K_4 and its planar imbedding.	186

Chapter 1

Introduction

You would probably expect a text book on a "pure" branch of pure Mathematics, such as Topology, to start with a definition of Topology and rapidly go on to an explanation of its contents with its theorems and their proofs. That is *not* the way we will present the subject. That would take the *fun* out of it. For one thing, you do not only want to know *what* you are doing but also *why ever should you be doing it?* For another, it makes the subject much more difficult to follow. Definitions, taken out of context, are not very enlightening in themselves. They must be seen in context to follow what they are all about. They must be clarified by examples and counterexamples. They must be "played about with" to see what would happen if they were modified.

Let us illustrate our point by presenting a short dialogue. Imagine, if you will, a parent defining the word "chair" for a child. The child (able to talk) has no idea of what the word means but has seen the actual objects that are called chairs.

Mother: A chair is something with four legs.
Son (pointing to their cat): You mean that.
Mother: No dear, not the cat but something you sit on.
Son (who has often tried to take short rides on his dog): Ah! You mean Rover.
Mother (patiently): No Son, not Rover. Something you can go on sitting on for long times.
Son (pleased to have followed at last): You mean that pony I rode.
Mother (less patiently): No, no! It is man-made.
Son (tiring of the guessing game points to the floor): Do you mean that?
Mother (definitely irritated by now): Why *are* you being so slow today! *(Pointing to a chair)* I mean *that*.

The parent would have been well advised to have started by pointing to various chairs and saying that they were chairs. This would, hopefully, provide the child with a feel for all those objects that would be regarded as chairs. However, it would probably go wrong by identifying other similar objects as chairs also. It would be necessary to point, also, to those types of objects and say that they were not chairs. The point is that it is not enough to define something, or even to explain (by examples) what it *is*; it is also essential to explain (by counterexamples) what it *is not* [Steen and Seebach, 1995].

In writing this book our hope is to explain what the subject is about and why it is worth understanding. For the latter purpose we hope to be able to convince you that it is fun and it is useful. As such, where concepts are introduced or results are proved, we will try to indicate how and where they are used. The "use" will not be in the sense that most lay people use the term with the addition "in everyday life". Rather, the use will be, for the most part, further in Topology; or in other parts of Mathematics; even (occasionally) in other fields. We will also play around with the definitions a little bit so that you can understand better *why* they are chosen as they are. Generally, they are the most convenient and simplest that *could* be chosen. We hope that you will learn to see Topology as a *simplification* which is useful rather than as an unnecessary excursion into abstruse erudition.

1.1 The Broad Context of Topology: Mathematics

To put Topology in context within the broad field of Mathematics it is first necessary to consider what Mathematics is. Of course, following our policy about coming out with definitions, we will not try to define it but to explain it. Essentially, Mathematics is a language — or perhaps a whole class of languages. It is a language in which the "words" used have only one meaning. This is not to say that a given symbol may not be used in two different places in totally different ways. One may often need to use the same symbols for some totally different purposes. However, the significance assigned to some symbol will be unique in a given discussion. Generally the attempt is made to avoid duplication of usage of symbols, but the number of symbols being limited (and the imagination of the users being further limited), all too often symbols are used in widely different senses. This double usage is analogous, for example, to the use of the word "light" as opposite of "dark" and also as opposite of "heavy". Nevertheless, in any given usage the mathematical "word" has only one meaning. This claim would not hold for other languages. For example, despite the claim that "a rose is a rose", it is not. Apart from the flower it denotes, the word is heavy with connotation. It carries with it the wealth of all poetic references to it — its use as a simile, as a metaphor, as a symbol. A "rose" is so much more than a rose.

Mathematics avoids connotations in the above sense. The price paid for this avoidance is that it cannot be used for poetic and purely "literary" purposes — it cannot be used to express and evoke emotional responses. That is the price – the price and the benefit! In losing emotional content it gains precision. There is no question of convincing people because they would *like* to believe what is said – or for that matter of putting up someone's back. There is no danger of inadvertently, or deliberately, confusing someone with double meanings. To avoid this possibility mathematicians generally start by specifying what each symbol being used stands for with statements like "Let X be so and so and x be such and such", which strike the non-mathematicians as being variously funny, pedantic or irritating. To conclude, a mathematical argument is there for all to see and judge without any ambiguity and without inducing any prejudices.

The language of Mathematics is not only precise, it is concise. This aspect of Mathematics is of supreme importance. A long argument in words may often lose and confuse one. By the end of it one has forgotten how it began. It is difficult to retain the argument in its entirety in one's "current memory" available for instant recall. However, when those pages are reduced to a few lines, the argument can be seen as a whole. Much of the development of Mathematics, particularly for the purposes of application, has been the introduction of new notation. New definitions open up new concepts and new notation makes them more readily applicable. As such Mathematics is the language for precise arguments, however long or complicated.

Another aspect of Mathematics, to which we will return later is that it is quantitative. One can, of course, deal with quantities in other languages (for example, in English). However, the usage is cumbersome there. Mathematics has been developed specifically with a view to maintaining convenience in quantitative reasoning. In this context one often thinks of Mathematics as a *tool*. In some sense all languages are tools for conveying ideas, and often feelings. However, other languages have simply evolved by the random forces of usage. Mathematics has developed in a directed way. If we liken an ordinary language to a pointed stick picked up by a primitive man for the purpose of hunting, Mathematics could be likened to a bow and arrow or a rifle.

Yet another aspect of Mathematics, and one of special relevance for us, is its ability to deal with spatial relationships. We will return to this aspect later as well. For the present we want to stress its importance for application to other branches of the human endeavour to understand and manipulate the environment. This ability is essential for building things that function as required. Thus Mathematics is a tool not only in a general quantitative sense but also in the very precise sense of spatial relationships.

In view of its precision, conciseness, special adaptation for quantitative discussion in general, and for spatial relationships in particular, and because of its overall adaptability, it is the language of science and engineering. It is for this reason that Mathematics has been extolled as "The Queen of all sciences" and "The Mother of all Sciences" — in a most un-mathematician-like way we might add. One cannot even claim that it is "the language of *all* sciences". It is, however, the language *par excellence* for Science. It is not, itself, a Science as it can not be falsified (as is requisite for a scientific theory), it is not tentative as Science *must be*. However, some part of it will practically always be used in most sciences and their technological applications.

1.2 The Development of Mathematics

We are not, here, going to trace out the entire history of mathematical development but will, rather, focus on those aspects relevant for understanding Topology. This is necessary to understand the context in which Topology can be placed and hence follow the topological way of thinking.

It is generally believed that Mathematics had its start in counting, which would have been necessary to develop the most primitive barter economy. As such it may well pre-date the development of other languages. Social cooperation would, presumably, have required primitive economic transactions. Discussions of this topic are available in books, of a popular or more scholarly nature, on the history of Mathematics. The popular books of Sir Lancelot Hogben, in particular his book "Mathematics for the Million"[Hogben, 1968], are particularly readable. (It may be mentioned that he discusses the grammar of Mathematics in some detail there.) Suffice it to say, here, that probably the earliest of the mathematical languages is *Arithmetic* — the language of counting. To this day there is confusion between Mathematics as a whole and its sub-branch *Arithmetic*. The specialization of applied Arithmetic, which itself has vast applications, is *Statistics*. It is now generally regarded as a study distinct from Mathematics.

A somewhat later development follows the invention of agriculture. As man turned from the nomadic life to a settled life it became necessary to demarcate land for exclusive agricultural use. This development led to the invention of *Geometry* — the language of spatial relationships. To some extent this subject was an application of Arithmetic as the spatial relationships were quantitative. However, where primitive Arithmetic used *natural numbers* (positive integers) for counting and all *integers* (including negative integers) for trade, Geometry required the introduction of *ratios* of these numbers to deal with lengths and angles that were not directly commensurate. Thus were born the *rational numbers*.

This type of interaction is typical of the development of Mathematics. First one branch develops then it leads to another. Now both branches develop together and each adds to the development of the other.

Whereas the beginnings of both Arithmetic and Geometry are lost in the mists of time, we know the subsequent development of counting systems with different bases by the very early civilizations in the Indo-Pak sub-continent, in Mesopotamia, in Egypt, by the Chinese and by the Mayans. Also, we know of the development of Geometry for the purposes of construction and land demarcation. However, the first signs of clear-cut *abstraction*, which is the hallmark of modern Mathematics, comes from the Greek civilization. To that extent Mathematics may be thought of as the invention of the Greeks. They developed, also, the concept of *rigor* for mathematical proofs. The model for proofs, to this day, is the geometrical proofs developed by the Greeks. The development of Geometry may be dated more precisely from just before the time of the Pythagoras, whose society's profound theorem stands at the base of so much of Mathematics. The entire system of Geometry may be thought of as reaching maturity with Euclid's compilation of the subject in a series of books.

The next major development comes from the Indo-Pak sub-continent – the numeral system that is generally credited (in English) to the Arabs is called "hindsa" meaning "from Hind". (There are claims that it came originally from China but they do not seem to be adequately substantiated.) The fundamental invention was the zero, which allowed a place value to be assigned to numbers, making it possible to conceive of a totality of numbers using a finite set of symbols, i.e., the digits zero to nine. Further, it allowed a simple procedure to do arbitrarily complicated additions and subsequently subtractions and multiplications. Subtractions are "reverse additions" as it were and multiplications "repeated additions". Treating multiplications as new, independent, operations, division arises as a "reverse multiplication". A much deeper appreciation of numbers and the arithmetical operations on them became possible with this new development of the numeral system. That was not possible, say, in Roman numerals.

The next real innovation came from the Arabs. It was the development of a generalized Arithmetic by Al Khwarizmi. There are claims that it originally came from the Indo-Pak subcontinent (or even from China) but these are far from substantiated. The new innovation was to represent an arbitrary given number by a letter of the Arabic script and use a procedure to give general results. Thus one got formulae like

$$(a+b)^2 = a^2 + 2ab + b^2 \ , \ (a+b)(a-b) = a^2 - b^2. \tag{1.1}$$

Further, it could be used to solve problems in which an unknown quantity appeared. In a book entitled "Al Kitab al Mukhtasar fi Hisab al Jabr wal Muqabala", Al Khwarizmi described a procedure for determining the unknown quantity. The title means "the (powerful) calculational procedure of transposition and comparison". Much later, when this book was translated into Latin, the translation began with "Algorism hath spoken that..." and went on to describe his powerful "Algebra". Thus was born *Algebra* and the term "algorithm" used so commonly in computing nowadays. It also gave birth to the term "logarithm", used for multiplying large numbers by the John Napier of Merchiston. Since ship's captains used a calculational procedure to work out where they were on the sea, by maintaining a diary of speed and direction of travel, the diary itself came to be called a "log book" and has hence led to the "captain's log" of the TV series "Star Trek" fame. But all this takes us very far from our main goal of Topology and we had better revert to the main track.

Algebra was developed apace by the Arabs and was used to solve various geometric problems as well. Al Khwarizmi had already shown how to deal with *quadratic equations*. Omar Khayyam, developed methods for solving *cubic* and *quartic* (bi-quadratic) equations. (It is ironic that this great scholar and innovator is now remembered only for his poetry through a translation into English by Edward Fitzgerald, and a name to be remembered at par with Pythagoras and Euclid is regarded as a mere source of inspiration for a not-so-major English poet.) But let us not dwell on the iniquities of history. Instead we proceed with our story.

Algebra led naturally to numbers that were not ratios. Solutions of quadratic equations required irrational numbers just as Geometry had required non-natural numbers. However, though the Greeks had equated *"irrational numbers"* with "insane numbers", they were no more "insane" than rational numbers are "unnatural". The terminology is an unfortunate legacy of the history of development of the number systems. The worst was yet to come. The Arabs could not believe in the solutions of equations which involved the square-root of a negative numbers. After all, both positive and negative numbers, when squared give a positive number. The above statement is typical of how the way one says something can block future developments and solution of problems. It is how ordinary language is inferior to Mathematics for the solution of problems. By saying "both" the Arabs excluded the possibility of anything else!

After the collapse of the Arab civilization and the *renaissance* (re-birth) of Western civilization it occurred to Gerolamo Cardano to provide a third alternative. Define a *new type of number* whose square is negative. Most unfortunately, this new number was called *imaginary*, as opposed to the other *"real"* numbers, and denoted by the Greek letter *iota* (now commonly written as the italicized Latin letter *"i"*). Of course, imaginary numbers are no more un-real than irrational numbers are insane and rational numbers are un-natural. Still, there is a very common prejudice against using these numbers to deal with *actual* physical quantities, purely because of the name. (Yet another example of how ordinary language distorts one's thinking.)

Analytic Geometry was soon extended from two to three dimensions. Here curves were studied first and then surfaces in three dimensions. With some help from others, Johann Carl Friedrich Gauss was principally responsible for the theory of surfaces. In particular, he showed how the geometry of surfaces could be studied internally, from within the surface, instead of having to enter the third dimension. Later Georg Friedrich Bernhard Riemann, and to some extent Nikolai Ivanovich Lobachevsky, extended the analysis to higher dimensional curved spaces of different kinds (see Figure 1.1).

Calculus and Geometry were extensively used in physical and engineering applications and hence were very thoroughly studied and developed. Riemannian geometry continued to be regarded as an abstruse bit of pure Mathematics with no practical relevance. However, with the advent of Einstein's theory of general relativity, it too became an applied field and was extensively explored and developed – particularly for four dimensions.

In our story of "The Rise and Rise of Mathematics" we have left out many important developments or whole new fields of Mathematics, such as the branches of "modern algebra". This is not because they are any less important than the branches mentioned by us, but because they were not relevant for the central "hero" of our story – Topology.

(At this stage imagine you are watching a film. Now comes the loud music and the title of the film Topology and the rest of the credits. Bear in mind the film "Star Wars". We have finally arrived, after our long detour, at our destination.)

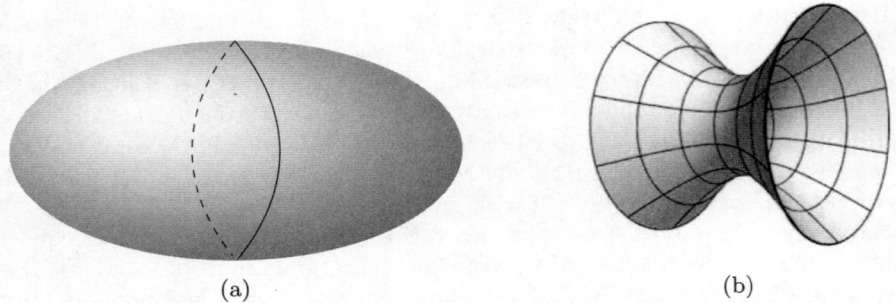

(a)   (b)

Figure 1.1: (a) The spheroid above is an example of Riemannian geometry. Notice that a triangle drawn on the surface of a spheroid will have the sum of all internal angles greater than 180°, unlike a triangle in Euclidean geometry. (b) A hyperboloid of one-sheet above is an example of Lobachevskian geometry. Unlike the Riemanninan and Euclidean geometries, a triangle drawn on the surface of the hyperboloid shown will have the sum of all internal angles less than 180°.

1.3 The Advent of Topology

Long long ago, far far away, in the seventeenth century, when the Earth was in attendance about the Sun roughly a fifth of a light year back in the galaxy, two noblemen are standing near a bridge in France. The view pans out and we see seven bridges connecting two islands in a river to the two shores and each other (see Figure 1.2). The view closes in again and we see the two talking.

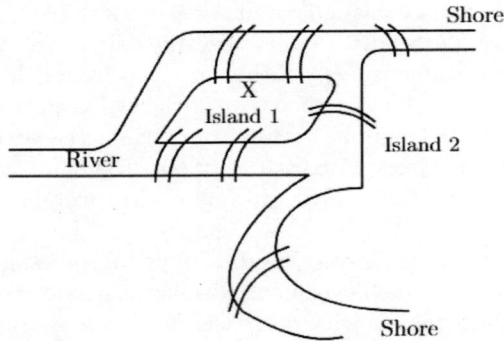

Figure 1.2: The seven bridges problem. There is no way to start from the point marked X and return to it crossing every bridge without recrossing any.

First Noble: But Armand, I tell you it is impossible – totally impossible – to start on one of the islands and return to it having crossed all seven bridges without recrossing one of them.
Armand (not questioning the importance or relevance of crossing all the bridges in this manner): But Henri, you just have not tried doing it the right way. Of course it *must* be possible.
Henri: I spent all night trying. Had it been possible I would surely have found it — wouldn't I?
Armand: Knowing you, Henri, I doubt that very much.
Henri (hotly): Are you saying I am a fool?

Armand (with a sneer on his face): No. I am saying you are a dunderhead and an idiot!
Henri (pulling off his glove and slapping Armand's face with it): I challenge you to prove that you can cross the bridges. *(He draws his sword.)*
Armand (drawing his own sword, salutes with it): I will.
They start duelling. The camera moves back and we see two not-so-noble men.— Not so noble, but a whole lot more sensible, one draws out a piece of paper, the other a pencil (forgive the anachronism here) and both start to draw Figure 1.2. Thus is Topology born!

The scene, though fictitious, contains half the truth (the other half being for effect). Topology *did* develop from the so-called "seven bridges of Königsberg problem", popular among some French nobles, though the bridges are not in France. Apparently, they *did* try solving it by doing it practically. It *was* finally resolved by the not-so noble, Swiss mathematician, Leonhard Euler in 1736 [Euler, 1736], using pen and paper and a method that also spawned Graph Theory (but that is another story). He proved that there is no way to achieve the single traversal required.

What is relevant for our purposes is that the sizes and shapes of the island, the bridges, the river and the shores are not important. All that matters is the *positional* relationships between these things. It does not matter whether the islands are square, rectangular or circular for that matter. Who cares whether the shores are irregular or made like a camel? Using pen and paper, instead of trying to do it practically, why bother if the bridges measure ten steps or a hundred? What matters if the river is deep or shallow? We would replace Figure 1.2 by Figure 1.3 and it makes no difference. *This* is the spirit of Topology.

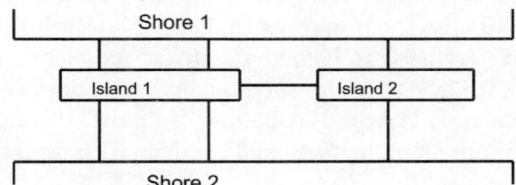

Figure 1.3: A topological representation of the seven bridges problem.

Another problem arose at about the same time [Stewart, 2010]. It was noticed that any map could be labelled by four numbers (or colored by four different colors) so that no adjacent regions have the same number (or color). By "adjacent" is meant that there are sides in common. Many regions could have one point in common, of course. For example, consider a circle cut into a vast number of "cake slices" through the center. They will all have the center in common (see Figure 1.4).

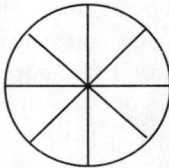

Figure 1.4: Many more than four regions can touch at a point. Here eight such are shown. The criterion for the four color problem is that there must be finite sides in common between the regions.

Of course, they would all require different numbers (or colors) if we had defined adjacent to mean that a point was in common. With the correct definition only four numbers (or colors) will ever be needed. (Try it and see, if you like.) This was known as "the four color problem" (see Figure 1.5).

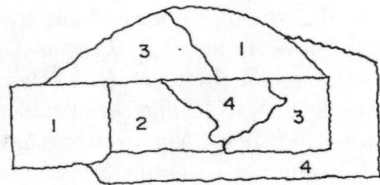

Figure 1.5: The four color problem. Every map can be colored using four colors so that adjacent areas have different colors. We use numbers instead of actual colors for this purpose here.

Again the shapes of the regions, or their sizes, are totally unimportant. Again, all that matters is their positional relationship.— Again we have the spirit of Topology.

As an aside, it may be mentioned that though this problem led to the birth of Topology, unlike Oedipus Rex, Topology did not manage to kill its father. The problem survived to give birth to *Graph Theory*. Till recently [Appel and Haken, 1977], that attempt was also unsuccessful. There has been a claim that it has been resolved using 1200 hours of computer time running an elimination procedure. Apparently, the proof was challenged. However, nobody followed the proof except the computer – and it refused to change its story. As such it is generally believed that Graph Theory *did* kill its father. (It could not be convicted, however, because of the judicial requirement that the evidence be "beyond a shadow of doubt". Though no one can shake the computer's story it is a very unreliable witness and mathematicians are still searching for a more believable proof.)

1.4 The Mathematical Background for Topology

As should have been clear by now, Topology is an extension of geometry that goes beyond the generalization achieved by Riemann or Lobachevsky. Let us elaborate on this extension further. In ordinary Euclidean geometry, the space in which we work is flat, and hence trivial. In the geometry of Lobachevsky the space remains flat but the theorem of Pythagoras does not hold. Instead of taking the sum of the squares of the difference of coordinates in each direction $((y_1 - x_1)^2 + (y_2 - x_2)^2 + \cdots)$ to define lengths, we would have sums and differences in general. Thus, for example, in a two-dimensional space, we could have the "length" defined by $\ell^2 = (y_1 - x_1)^2 - (y_2 - x_2)^2$, or in a three-dimensional space $\ell^2 = (y_1 - x_1)^2 - (y_2 - x_2)^2 - (y_3 - x_3)^2$ and so on. In Riemannian geometry we again lose Pythagoras' theorem. Here the square of the length remains a sum of squares but the coefficients of these squares need not be unity. Thus, for example, $\ell^2 = \alpha_1(y_1 - x_1)^2 + \alpha_2(y_2 - x_2)^2$ with $\alpha_1, \alpha_2 > 0$, but they are functions of position in general. The latter geometry becomes necessary when describing a curved space *internally*. For example, the surface of a sphere must be a two-dimensional space and so points should be identifiable by two numbers (the coordinates of the point). However, it is not generally possible to express the distance between two points as the square root of the sum of the squares of coordinate differences. It may work for two *specific points* on a *specific sphere* with a *specific choice of coordinates*, but it will not work in general. The Lobachevskian geometry has the *very* odd feature that the "length" squared need not be a positive quantity. Generally, for two distinct points, $\ell^2 \gtreqless 0$.

Riemannian and Lobachevskian geometries were not developed for any specific application but were mere intellectual curiosities. However, soon after the Albert Einstein published his 1905 paper on Special Relativity his teacher, the Hermann Minkowski, re-expressed the theory in geometrical form using a four-dimensional Lobachevskian geometry, see for example, [Qadir, 2020]. This innovation did not please Einstein in the beginning. However, as Einstein's attempts to generalize his theory to deal with arbitrary motion ran into difficulties he started seriously trying to learn more about Riemannian geometry from his friend, the Marcel Grossman. Finally, in 1916, he used a mixture of the two geometries to formulate his general theory of relativity. The mixture used by him is now called *pseudo–Riemannain* geometry. The immense and overwhelming success of the theory has made the new geometry almost a branch of Applied Mathematics.

Topology also dispenses with Pythagoras' theorem. It does not do so by modifying the definition of length but by ignoring it. After all, if sizes and shapes are not going to be important, lengths will be quite irrelevant. We could modify the definition practically arbitrarily. Dethroning the specific definition from its central role is a profound extension of geometry and the extension can well be regarded as a new subject in itself.

Let us not give the impression that lengths cannot be defined by Topology. Quite to the contrary, for a given space there may be many different measures of length definable. They are called *metrics*, from the Greek word for "measure" (μητρον, which in English is written as "meter"). Topology, in fact, provides for a change of definitions of length — a change of metrics. There may be spaces on which no concept of length applies. These are called *non–metrizable* spaces. A space with a length measure defined on it is called a *metric space*. To provide examples of these spaces it is necessary to explain what a "space" signifies to a mathematician.

In general a space is a set of elements, which are called *points*. Further, there is generally some *structure* implicit in the space arising from the context of its use. The set may or may not be finite. Consider the set of points on a sheet of paper (idealized as having no thickness) as our space. This together with a meter scale is an example of a metric space. For the non–metrizable space we would like to present an example that needs a slight background of economics. In economics people are generally thought of purely in their economic roles. One is not concerned with whether they are short or tall, fat or thin, happy or unhappy. All that matters is how they interact with each other economically. Broadly speaking, they may be thought of as having two major roles. They may be *producers* who produce goods or *consumers* who use them. Producers are generally further subdivided. They may be labour, who are directly involved in production, *managers* who organize labour to work together smoothly and entrepreneurs who provide the capital outlay to make the goods. Not everybody would be a producer and a given producer may play more than one role in the production process. However, everybody must be a consumer. Consider the set of all consumers in the economy as our space. We could provide them all with ID numbers. However, these numbers would merely be *labels,* they would have no economic significance. Interchanging ID numbers would not change the economic reality being described. This is an example of a non–metrizable space. Though we do have numbers here, they do not provide a length measure.

The arbitrariness of the definition of length is crucial for the topological approach to a problem, where lengths are unimportant. Remember that the usual definition of the length of a vector with the help of Pythagoras' theorem is essentially built into the definition of the "dot product" of two vectors. In fact for any vector **a** we can write the length squared as $a^2 = \mathbf{a} \cdot \mathbf{a}$. Since $\mathbf{a} \cdot \mathbf{b} = ab\cos\theta$, where a represents the length of **a** and b the length of **b**, while θ is the angle between the two; we use this "dot product" to define the direction of one vector relative to another (see Figure 1.6). By changing the definition of *length*, then, we would also change the *shape* described as all directions as well as sizes would be altered. This change may be visualized by thinking of a geometrical figure drawn on a rubber sheet which can be stretched and squashed differently in different directions thus distorting the original shape. In fact, for this reason Topology is often described as "rubber-sheet-geometry".

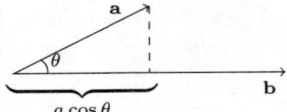

Figure 1.6: The component of a vector **a** along **b** is $a\cos\theta$.

One could equally well think of figures that are closed as made of rubber bands. Without twisting the band we could change its shape so that it is in the form of a triangle, a rectangle, a more arbitrarily shaped quadrilateral, a pentagon, or for that matter an oval or a circle, etc. So long as the band is not cut or twisted over itself, it remains very much the same despite its changes of shape. This is the spirit of Topology. We regard the rubber band as *topologically* the same even though it does not remain *geometrically* the same. The subject of Topology attempts to understand, define and use this concept of invariance. To explain by analogy, in Geometry we started with theorems of *congruent* triangles (and other shapes) and extended to the concept and theorems for similar triangles (and other shapes). In this sense Topology takes the next step of generalization. We now regard all such shapes that divide space into two parts, an *inside* and an *outside*, as topologically equivalent.

1.5 A Psychological Aside

Recent developments in psychology have some bearing on our subject of discussion. We would like to, here, impose on your patience a bit further and explain the relevant development in our own roundabout way.

An odd feature of the way our brains work is that whereas the eyes and ears connect directly to the brain, all other nerves cross over. The cerebrum consists of two "hemispheres" — the left and the right. The eyes and ears are connected left to left and right to right. However, the hands are connected left to right and right to left. Of course the two hemispheres are also connected to each other making the brain capable of acting as a whole.

While studying brain damaged patients it was found that occasionally the two hemispheres got disconnected. These patients were used to study how each hemisphere functioned. The method adopted was to close one eye and set the patient an activity to do. If the left eye was closed and the activity was required of the *right* hand, the patient failed to perform it. However, with the *left* hand it would be performed. Similarly, input from the *left eye* evoked response from the *right hand*. Using this fact different responses could be evoked from each hemisphere separately. Though not entirely, it appears that the two halves of the brain have separate, specialized, functions. The left side seems to perform all analytical functions such as arithmetic, symbol manipulation, logic, language, etc. The right side seems to be where aesthetics, intuition and other synthetic functions reside. It helps to see things *as a whole* rather than to take them apart.

In the context of Mathematics, Arithmetic and Algebra are clearly left hemisphere activities. However, Geometry may be much more a right hemisphere activity. It seems to have the same type of aesthetic appeal and approaches problems as a whole rather than as its parts. Topology, being an extension of Geometry, presumably also falls into the right hemisphere.

Another brief aside is that the human brain, unlike that of the dog or the porpoise, is essentially an enlargement of the visual part of the central nervous system rather than the olfactory or the audial parts. Hence our greater appeal to and reliance on seeing, even identifying it with understanding. This fact brings out the importance of the *geometrical*, and by extension, *topological* approach.

1.6 Topological Constructions

As with Geometry, Topology is based on constructions. The formal procedures will come later. Here we would like to present the simplest types of topological construction, whereby one topological shape is changed into another.

The very simplest is to consider an elastic string and join the two ends together. The result is an elastic band. Conversely, cut an elastic band and we obtain an elastic string. The two are fundamentally different. For example, we could meaningfully assign an order to the points on the string. There would be some points coming before others and some coming after. One end would be before all other points while the other would be after all others. However, this would not apply to the band. If one point was prior to another, that point would be prior to a third which would be prior to the first, see Figure 1.7a. Again, consider two points on each of them. For the string there is a unique path from one to the other, while for the band there are two distinct paths, see Figure 1.7b. Further, if the string is cut across it will break into two pieces while the band, when cut, will break into the one string, see Figure 1.7c.

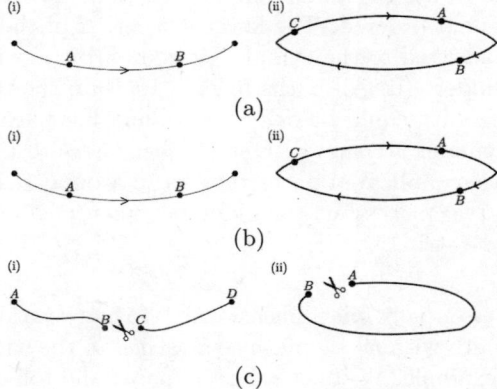

Figure 1.7: (a) A string of elastic is shown in (i) and a band in (ii). The string has two ends, the band none. Points on the string are well ordered, in that we can define A as preceding B. However, for the band, since B precedes C and C precedes A, B precedes and succeeds A. Thus there is no good ordering available. (b) There is only one path from A to B on the string but two distinct paths on the band. (c) On cutting the string we get four ends in all and two separate strings. Cutting the band gives only one.

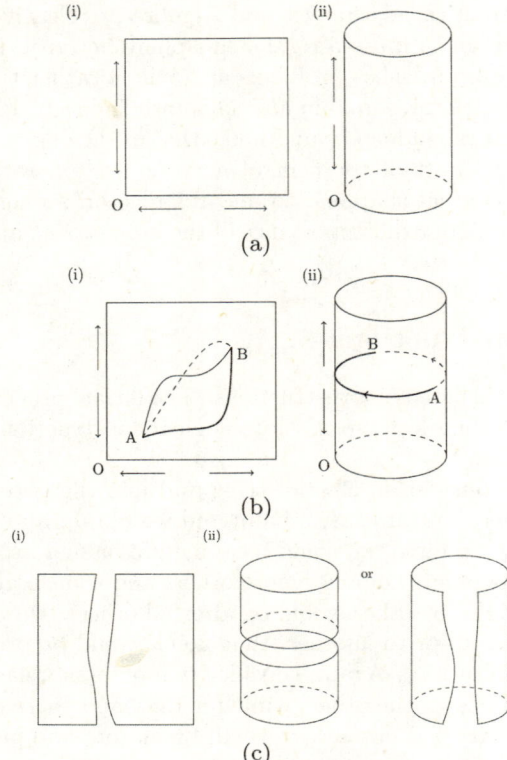

Figure 1.8: (a) The rectangle (i) has each direction ordered while the cylinder (ii) has one direction (around the cylinder) that is not ordered. The lower left corner of the rectangle in (i) is labelled O. Notice that when it is converted to the cylinder, it is an arbitrary point on the circumference of the base circle of the cylinder. (b) All paths from A to B on the rectangle are equivalent, in that a string lying along one path could be deformed to another path without breaking it. Two inequivalent paths on the cylinder are shown. One (dashed) goes to the right, whereas the other (solid) goes to the left. More complicated paths that wrap around the cylinder are also possible. (c) The sheet gets cut into two pieces but the cylinder, when cut, may be cut into one or two pieces — two cylinders or one strip.

The above example was essentially one–dimensional. We had regarded the band and the string really as providing only a length with no significance attached to the width and the thickness. Now consider a two–dimensional example. We take a sheet of paper and roll it into a cylinder, gluing the opposite edges together. Again the two objects are topologically distinct. On the sheet, starting at one corner (say O) each direction can be ordered, as depicted in Figure 1.8(a)(i). On the cylinder there is certainly an up and down ordering) but no ordering around, see Figure 1.8(a)(ii). Cutting the sheet would give two pieces, see Figure 1.8(c)(i), but for the cylinder there are additional complications. If we cut *around* we get two cylinders, while if we cut from *down to up* we get a single sheet (see Figure 1.8(c)(ii)). Also, on the sheet, though we have infinitely many paths possible between two points they are *topologically* the same. Imagine the two points A and B as pins holding down a thread, see Figure 1.8(b)(i). If the thread is elastic, it can be deformed about from one path to another without cutting it or tying it into knots. This does not apply to the cylinder. We can have many topologically inequivalent paths, see Figure 1.8(b)(ii). One could classify the paths according as whether they go to the left or right and whether they wind around the cylinder or not. If they do so, how many times they wind around determines their class.

A new twist can be added – and we do literally mean a *twist* — in the above example. For the sheet, take it to be sufficiently narrow to be taken as a strip. Now, instead of joining the opposite edges together directly, first twist the strip and *now* join the opposite ends together. (See Figure 1.9a.) This object is not equivalent to either the strip or the cylinder. For one thing, both the sheet and the cylinder have two faces. It can be verified by running a pen right round the strip that it has only one face. Further, if one puts arrows going up where one starts they will point down when we get to "the other side" of that point, see Figure 1.9b. If we cut along the length of this object we find that we get a single doubly twisted cylinder instead of two cylinders, see Figure 1.9c. This object is called a *Möbius strip* [Stewart, 2010]. It is clear that it has *very* different properties from the strip or the cylinder. The 3 dimensional analogue of the Möbius strip is known as *Klein bottle* and is displayed in Figure 1.10.

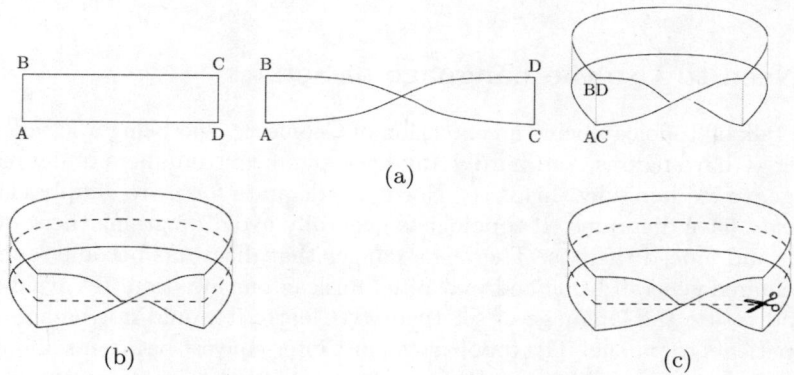

Figure 1.9: (a) Take the strip $ABCD$ and twist it, then join the opposite edges AB and CD so that C lies on A and D on B. This is called a Möbius strip.
(b) Draw a line around the Möbius strip keeping the pen on the paper at all times. The line drawn is a closed curve which passes on both sides of the original strip. As such the Möbius strip has only one side. (c) Cut the strip along the line drawn. This does not cut it into two pieces but leaves a single doubly twisted cylinder. The secret of its "unbreakableness" is the twist which joins the upper part to the lower part so that the two parts do not separate.

Yet another object can be constructed from the strip or sheet. Make a long tube (essentially a cylinder) and join the two ends to get a tyre shaped object (or a doughnut shaped object if you prefer eating), see Figure 1.11. This is called a *torus*. If we had the sheet made of rubber we could have pulled the edges round so that all corners would meet at a point and then seal all the edges and the corners together to get a ball, or more properly a "sphere" in a topological sense. It is important to bear in mind that the topological "sphere" is different from the usual (geometrical) sphere which has a constant radius. This is a stretchable–squashable–squeezable sphere.

Figure 1.10: A Klein bottle is a 3–d version of the non-orientable surface, which has only one side. It can be constructed by bending two Möbius strips and pasting them together as shown here. In this case, despite its appearance, there is no inside of the bottle — it can hold nothing in.

Figure 1.11: A torus is like a tyre tube or a doughnut. Imagine two rings interlocked and held orthogonal to each other. Now one ring is run around the other. The resulting surface is a torus. This procedure is the same as taking the Cartesian product of one circle with the other, $\mathbb{S}^1 \times \mathbb{S}^1$.

1.7 The Need to Use the Language of Sets

Despite all our talk of Topology being an extension of Geometry, and being a *visual* study, few text books of Topology have figures, and hardly any have significant numbers of figures. It behooves us, at this stage, to explain why this is so. Nor is it adequate for us to simply claim that other text books *should* have diagrams. If topologists generally avoid diagrams there must be a good reason for it — and indeed there is. There is a danger that diagrams become too geometrical to allow for the required generalization and may often mislead our intuition. To avoid this possibility topologists tend to use the language of set theory. (Clearly it would not be a good subject for brain damage patients to pursue. The topologists must inter-convert between symbol manipulation and geometric, wholistic, visualization.) Before going on to the promised review of set theory, we would like to give an example of how the use of diagrammatic thinking may confuse us.

It seems obvious that a finite area would always be bounded by a finite perimeter. One could try to verify this belief by constructing diagrams. Since we would be unable to construct an infinite perimeter bounding a finite area we may conclude that we have proved that is impossible. This proof would, however, be wrong. Consider a rectangle with sides a and b. The area would be $A = ab$ and the perimeter $2(a + b)$. Holding A fixed we could increase a and decrease b so that $b = A/a$. Thus the perimeter will be $2(a + A/a)$. Clearly this can increase indefinitely for a given A. Hence the limiting figure (which we cannot draw) has a finite area but infinite perimeter (see Figure 1.12).

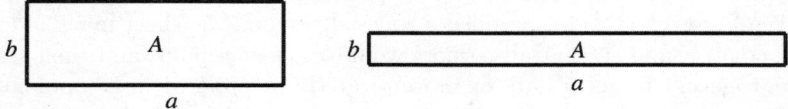

Figure 1.12: Increasing the length and decreasing the breadth of a rectangle can leave the area fixed while increasing the perimeter indefinitely.

A more interesting example is provided by the following construction. Start with an equilateral triangle of sides a. Now, centered at each vertex construct equilateral triangles of side length $a/3$ and delete the interior lines (see Figure 1.13). Repeating this process indefinitely we again obtain a figure with finite area and infinite perimeter. (This object is an example of a *fractal*, i.e. a "fractional dimensional" figure. The crucial feature of a fractal is that it is self–similar in that if we focus on a small part of it, it looks just the same as the larger scale object does. Such objects are of great interest in the field of *branching theory* or bifurcation theory, more popularly known as *chaos theory*. We will not be dealing with these topics, or such objects, here.)

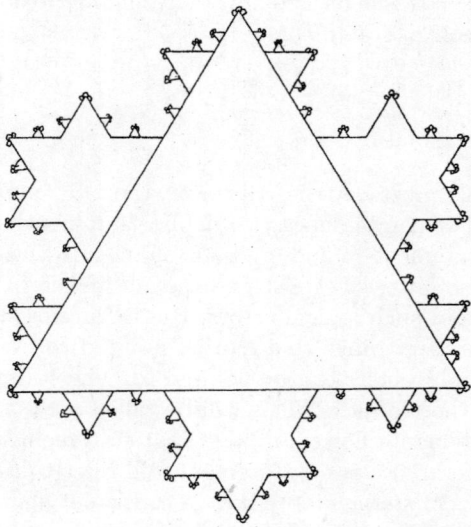

Figure 1.13: A more interesting example of a figure with a finite area and an infinite perimeter. At all scales it looks the same. Such objects are called fractals. The resulting figure is called a Koch curve.

Unlike most other authors of Topology texts, we will not abjure diagrams. Intuition needs visualization and an introductory text must appeal to it. We, therefore, only warn the reader of the dangers of relying on diagrams and then continue to use them, while taking care to avoid being misled by them.

1.8 A Diversion into Logic

A *set* is a collection of objects. Of course, this definition is like most, circular. The objects, called *elements* of the set, are defined in terms of the set. However, we *know* what we mean. A word of caution is in order here. There is a risk of putting logically different things together in a set. For example the set of chairs in a room is well enough defined. However, we could not include in that set the colors in the rainbow or the numbers less than 10. To appreciate the problem we need to briefly discuss the problems that can arise if one does not take adequate care. These are called logical *paradoxes*.

A paradox is a set of statements which cannot be consistently taken to be true or false. There are many that are only apparently paradoxical, arising from confusion about the fundamental rules of logic (most often by confusing the concept of "implication" with "equivalence" of statements) and occasionally by subtle changes of the meaning of words. We are concerned with more definite paradoxes here. A famous paradox is based in Greek times. The people of Crete (called Cretans) were supposedly famous as liars. The paradox has it that a Cretan says "All Cretans always lie". If this statement is assumed to be true, the Cretan saying it must be lying. Hence it must be false! Conversely, if it is assumed to be false, then the Cretan was telling the truth and hence it is true!

The Bertrand Russell resolved this paradox by analyzing it [Doxiadis et al., 2009]. In fact it is equivalent to the following pair of statements:

1. The next statement is true.

2. The previous statement is false.

Assuming "1" to be true, "2" must be true. However, then "1" must be false. Thus "1" is true implies that "1" is false. (For completeness we define "implies that" which is denoted by the symbol "$\implies$". Given two statements p and q, "$p \implies q$" means that if p is true q must be true.) Again if we assume "1" to be false, "2" must be false and hence "1" must be true. Thus "1" is false $\implies$ "1" is true. To avoid such logical errors, Russel proposed his theory of classes. Start with concrete objects. These are things that can be pointed to. We can now make statements about them. Thus, for example, chairs can be pointed to. The statement "chairs are made to sit upon" is a statement about those objects. This will be called a *class 1 statement*. We could make a statement about that statement. For example "That statement is not entirely true". This is a *class 2* statement. In general, a class n statement will refer to a class $(n-1)$ statement. By defining objects to be *class zero statments* we have a consistent definition. By this procedure the above pair of statements is legislated out. If the first is class n, the second statement must be class $(n-1)$ and hence can refer only to class $(n-2)$ statements and not to the class n statement above.

The above theory manages to "explain" various other paradoxes. Another famous paradox, due to Russell, is the question "Does the catalogue of the library refer to itself?" Since the purpose of the catalogue is to find books, it makes no sense for it to refer to itself. Of course, regarded as a book it could be referred to, but *in its capacity as a catalogue* it is a class 1 statement about the books in the library and will not refer to itself as such. Another example is the following paradox. "A male barber in a town shaves all those men and only those men who do not shave themselves. Who shaves the barber?" The problem is that if the barber shaves himself, he is one of those who shave themselves. Thus the barber does not shave him. So he cannot shave himself. However, if he does not do so he is one of those who do not shave themselves and so the barber shaves him. The paradox is again resolved by the theory of classes. The barber, as an individual in the town *can* shave himself. *In his capacity as a barber* he acts on the individuals in the town and is not an individual of the town. Clearly, if he goes to his shop, uses the facilities of the shop, charges himself and declares his income to be taxed, he does not shave himself as an individual shaving himself, but only as the barber. On the other hand, if he shaves himself at home, not using his professional facilities and not declaring the shave for taxation he shaves himself as an individual and not in his capacity as a barber.

What does all this have to do with sets? Well, it teaches the valuable lesson that we cannot mix statements of different classes without running the risk of falling prey to logical paradoxes. Thus, when we talk of sets as collections of elements, we need to bear in mind that they must all be of the same class.

1.9 The Language of Sets

Though we are sure that you are familiar with the language of sets we include a brief review for completeness. It will also be useful to provide the notation as we use it. A set, A, consisting of some elements $a, b, \cdots, n$ will be written as $A = \{a, b, \cdots, n\}$. We then write, for example "a belongs to A" as "$a \in A$". Another way of writing a set is by specifying some property that the elements must satisfy. A straightforward tautology should help clarify the notation. "A is the set of all 'a' such that 'a' belongs to A" would be written as "$A = \{a | a \in A\}$". A more meaningful example is the set of all counting numbers between ten and a hundred: "$A = \{n | 10 \leq n \leq 100\}$".

The *union* of two sets is defined as the set which contains all elements belonging to either of the sets and only those elements. Thus if C is the union of A and B

$$C = A \cup B = \{c | c \in A \text{ or } c \in B\}. \tag{1.2}$$

The *intersection* of two sets is defined as the set consisting of those elements which belong to both sets.
$$C = A \cap B = \{c | c \in A \text{ and } c \in B\}. \tag{1.3}$$

The *set difference* is defined as the set consisting of those elements of the first set which do *not* lie in the second,

$$C = A \setminus B = \{c | c \in A \text{ and } c \notin B\}, \tag{1.4}$$

where "$\notin$" stands for the negation of "$\in$" and is read as "c does not belong to B".

Now consider two *disjoint* sets A and B, by which we mean that they have no element in common. In other words there is no element in their intersection. For consistency of our definition we *need*, nevertheless, to talk of the intersection as a set. We define this *empty set*, ϕ, also called the *null set*, to be a set without elements. This may seem like a contradiction in terms but is merely a convenient extension of the terminology of sets to enable a consistent definition of intersection. There is a common example of such an extension of definitions and terminology. You are, no doubt, familiar with the number of combinations of r out of n objects, written as ^nC_r or $\binom{n}{r}$. The expression for this quantity is $n!/[r!(n-r)!]$, where $n! = n \times (n-1) \times \cdots \times 3 \times 2 \times 1$ is the usual "n factorial". Now, obviously by definition, $^nC_n = 1$ as there is only one way of taking n out of n objects, namely take *all* of them. Thus $n!/[n!(n-n)!] = 1/0! = 1$. Hence $0! = 1$ for consistency. Notice that $0!$ was undefined according to the original meaning assigned to factorials. However, it is a convenient extension of notation. The null set is a similar extension.

Another commonly used term which may cause confusion is the universal set, normally denoted by U or X. We shall use the latter. The name seems to suggest a "universality" not present in the concept. It is merely the largest set under consideration. A subset of a given set is a set of elements contained in the given set, but not necessarily all of those elements. Thus if A is a subset of B, we write it as $A \subseteq B$ (also read as "A is contained in B") and we would then have, "$a \in A \Rightarrow a \in B$", which is read as "$a$ belongs to A implies that a belongs to B". This statement is equivalent to $B \supseteq A$ (read as "B contains A"). Two sets are said to be *equal*, $A = B$, if they have the same elements. Clearly, if $A = B$ then $A \subseteq B$ and $B \subseteq A$. We say that "A is a proper subset of B", if $A \subseteq B$ and $A \neq B$. In fact, for a proper subset, we also require that $A \neq \phi$. We define the *complement* of A as $A' = X \setminus A$ or $A' = \{b | b \notin A\}$.

For completeness we include some further notation that will be used later. Whenever we want to require that a given property hold "for all elements, x, of a set A", we will write it as "$\forall x \in A$". The phrase "such that" will be represented by "s.t.". The phrase "there exists an x belonging to A" will be written as "$\exists\, x \in A$" while if we want to signify that the x is unique we will write it as "$\exists\, !x \in A$". Thus, for example, the conditions for a group could be given as follows. If G is a set with a binary operation $*$ to be defined on it, $(G, *)$ will be a *groupoid* if $\forall a, b \in G, a * b \in G$. It will be a *semigroup* if $\forall a, b, c \in G, a * (b * c) = (a * b) * c$. It will be a *monoid* if $\exists\, !e \in G$ s.t. $a * e = e * a = a \forall a \in G$. It will be a *group* if $\forall x \in G, \exists\, !x^{-1} \in G$ s.t. $x * x^{-1} = x^{-1} * x = e$. The extreme brevity of the statements in this notation would be apparent to all. The corresponding ease of "seeing what is going on at a glance" would only become clear once there is sufficient fluency in use of the notation.

The matter of "implication" needs further discussion. When we say "$p \Rightarrow q$" we mean that "if p is true then q is true". However, it says nothing about p being false. In that case q could be true or false. Notice that "$p \Rightarrow q$" $\not\Rightarrow$ "$q \Rightarrow p$", where "$\not\Rightarrow$" stands for "does not imply", but rather "$p \Rightarrow q$" $\Rightarrow$ "$\not{q} \Rightarrow \not{p}$" where "$\not{p}$" stands for the negation of p. If "$p \Rightarrow q$ and $q \Rightarrow p$" we say that "p and q are equivalent" written symbolically as "$p \Leftrightarrow q$". This is also called a two way implication. An equivalent phrase to "$\Leftrightarrow$" with slightly different usage is "iff" which stands for "if and only if".

1.10 Counting the Elements of a Set: Cardinal Numbers

When we *count* we are actually setting up a one to one correspondence between some objects and a subset of the *counting numbers*. There are other, more formal, ways of defining counting numbers (such as the Lambda Calculus invented by Alonzo Church [Michaelson, 2013]), which were developed in the attempt to axiomatize all of Mathematics. When it became clear that the axiomatization approach would not succeed, the axiomatic definitions lost much of their significance. We will not, therefore, bother with trying to *define* counting numbers but merely take them for granted.

Along with the counting numbers came the concept of *adding* them. This concept comes most clearly from operations on the "number ray". Addition means moving that many steps to the right on the number ray. When numbers were used for extending bartering to *accounting* for barter transactions, the *inverse operation* of *subtraction* had to be introduced. With subtraction it became necessary to extend the counting numbers to include *zero* and get what are nowadays called the *natural numbers*. The reason is that if we consider $A \subseteq B$ and A has m elements while B has n, then $B \setminus A$ has $n - m$ elements. Clearly, then $A \setminus A = \phi$ has zero elements. We denote the set of natural numbers by $\mathbb{N}$.

For finite sets the number of elements in the set is, then, an element of $\mathbb{N}$. This is called the *cardinal number* of the set. Thus, if A has n elements we write it as $|A| = n$. The question then arises "what is $|\mathbb{N}|$?" This is not a trivial question.

Clearly, there can be no $n \in \mathbb{N}$ which is $|\mathbb{N}|$, for if we supposed that it were, we could always add some given number, say 1 to n and obtain a bigger counting number. This would have to, then, be taken as the new value for $|\mathbb{N}|$. The same argument would apply repeatedly. Hence we cannot assign a counting number to $|\mathbb{N}|$. We call such sets infinite. To give this a name we write it as $\aleph_0$ (read "aleph null"). Thus $|\mathbb{N}| = \aleph_0$.

We could have thought of $\mathbb{N}_1 (= \mathbb{N} \setminus \{0\})$ as the set that is generated from the element of 1 and the binary operation of addition if we require that our set be closed. (Of course, the number zero is not obtained from here.) Thus $(\mathbb{N}, +)$ is a *groupoid* (a set closed under a binary operation, as defined on p. 18). We now introduce the inverse binary operation of subtraction and find that we no longer have a groupoid (though it does provide the zero). To obtain a set for accounting we must include negative numbers so that it is closed under subtraction. This set is called the set of *integers* and is denoted by $\mathbb{Z}$. Thus $(\mathbb{Z}, -)$ is a groupoid. Of course, $(\mathbb{Z}, +)$ is also a groupoid. $(\mathbb{Z}, +)$ has the additional property of *associativity*, i.e., if $a, b, c, \in \mathbb{Z}, a + (b + c) = (a + b) + c$. This is not true for $(\mathbb{Z}, -)$ as $a - (b - c) = (a - b) + c \neq (a - b) - c$ if $c \neq 0$. We call an associative groupoid a *semigroup*. Thus $(\mathbb{Z}, +)$ is a semigroup. A semigroup, like $(\mathbb{N}, +)$, which has an *identity element* in it (i.e., $0 \in \mathbb{N}$ s.t. $0 + n = n + 0 = n \forall n \in \mathbb{N}$) is called a *monoid*. A monoid with a unique inverse for every element is called a *group*. Clearly, $(\mathbb{Z}, +)$ is a group while $(\mathbb{N}, +)$ is not.

Repeated additions were introduced into Arithmetic very early on, i.e., for $n \in \mathbb{N}, n + n = 2 \times n$, $n + n + n = 3 \times n$ and so on. This repeated addition is *multiplication*. Clearly, $(\mathbb{N}, \times)$ is a semigroup, as is $(\mathbb{Z}, \times)$, and in fact both are monoids. But they are not groups. The problem lies with the inverse elements under multiplication. Now, as with subtraction, we can define the inverse operation to multiplication, namely *division*, denoted by $\div$. Thus $(\mathbb{N}, \div)$ is not a groupoid. We define the groupoid generated by $\div$ to be $\mathbb{Q}^+$, the set of *rationals*. More correctly these are the positive rationals $\mathbb{Q}^+ = \{m/n | m, n \in \mathbb{N}; m, n \neq 0\}$. Clearly, $(\mathbb{Q}^+, \div)$ is a groupoid. Further $(\mathbb{Q}^+, \times)$ is a group. In fact, we can define the full set of rationals, $\mathbb{Q}$, by $\mathbb{Q} = \{m/n | m, n \in \mathbb{Z}; n \neq 0\}$ and $(\mathbb{Q} \setminus \{0\}, \times)$ is a group. Here $\{0\}$ is the set consisting of the element zero, of the set of integers. As such it is *not* the null set, which has no elements, but a set consisting of one integer, namely zero.

We can now ask for a comparison between these various infinite sets: $\mathbb{N}, \mathbb{Z}, \mathbb{Q}$, as regards their cardinal numbers. It would appear intuitively obvious that since $\mathbb{N} \subsetneq \mathbb{Z} \subsetneq \mathbb{Q}, |\mathbb{N}| < |\mathbb{Z}| < |\mathbb{Q}|$, where "<" stands for "strictly less than" as distinct from "$\leq$", which is "less than or equal to". However, this is one of the places where intuition is likely to lead one astray. To be able to compare cardinal numbers we need to go back to the definition of counting.

Consider, first, the set $\mathbb{M} = \{n | n \in \mathbb{N} \text{ and } n \geq 10\}$. Thus, we have $\mathbb{N} = \mathbb{M} \cup \{0, 1, 2, \cdots, 8, 9\}$. Now, generally, if $A = B \cup C$ then

$$|A| = |B| + |C| - |B \cap C|. \tag{1.5}$$

The last term is required to avoid double counting the elements that belong to both B and C. In our case, clearly there is no intersection between $\mathbb{M}$ and $\{0, 1, \cdots, 9\}$. Thus we have $|\mathbb{N}| = |\mathbb{M}| + 10$. However, we can construct a one to one correspondence between $\mathbb{N}$ and $\mathbb{M}$, as follows:

$$\mathbb{M} = \{10, 11, 12, \ldots, n + 10, \ldots\}, \tag{1.6}$$
$$\mathbb{N} = \{0, 1, 2, \ldots, n, \ldots\}, \tag{1.7}$$

i.e., generally if $m \in \mathbb{M}$, putting $m = n + 10$ we get a unique $n \in \mathbb{N}$ and conversely, given $n \in \mathbb{N}$ we can get a unique $m = n + 10 \in \mathbb{M}$. Hence $|\mathbb{M}| = |\mathbb{N}| = \aleph_0$. Thus, $\aleph_0 + 10 = \aleph_0$ and more generally, $\aleph_0 + n = \aleph_0$. Despite the fact that $\mathbb{M} \subsetneq \mathbb{N}$ the cardinal numbers are equal! This is not quite so surprising as it seems at first. We know that one plus one is *very* different from one, one plus ten is less different from ten, one plus a hundred is *only* 1% off the hundred, one added to a million can largely be ignored, to a billion is still more negligible and so on. Compared to infinity, one (or any other finite number) is *totally* negligible.

Now consider $\mathbb{N}_e = \{2n | n \in \mathbb{N}\}$ and $\mathbb{N}_o = \{2n+1 | n \in \mathbb{N}\}$. Clearly $\mathbb{N} = \mathbb{N}_e \cup \mathbb{N}_o$, $\mathbb{N}_e \cap \mathbb{N}_o = \phi$ and $\mathbb{N}_e, \mathbb{N}_o \subsetneq \mathbb{N}$. Thus $|\mathbb{N}| = |\mathbb{N}_e| + |\mathbb{N}_o|$. Again, there is a one to one correspondence between the sets as

$$\mathbb{N} = \{0, 1, 2, \cdots, n, \cdots\}, \tag{1.8}$$
$$\mathbb{N}_o = \{1, 3, 5, \cdots, 2n+1, \cdots\}, \tag{1.9}$$
$$\mathbb{N}_e = \{0, 2, 4, \cdots, 2n, \cdots\}. \tag{1.10}$$

To every $n \in \mathbb{N}$ there is a $(2n+1) \in \mathbb{N}_o$ and $2n \in \mathbb{N}_e$ and *vice versa*. Hence $|\mathbb{N}_o| = |\mathbb{N}_e| = \aleph_0$. Thus $2\aleph_0 = \aleph_0$ and more generally $n\aleph_0 = \aleph_0$. We now note that when we go to very large numbers it no longer matters how many times that large number we take. If someone is a billionaire who cares whether it is 2 or 3 billions that are owned! For an "infinitaire" it becomes *totally* irrelevant.

Since $\mathbb{Z} = \mathbb{N} \cup \mathbb{N}_-$, where $\mathbb{N}_- = \{n | -n \in \mathbb{N}\}$ and $\mathbb{N} \cap \mathbb{N}_- = \{0\}$, we have $|\mathbb{Z}| = |\mathbb{N}| + |\mathbb{N}_-| - |\{0\}| = 2\aleph_0 - 1 = \aleph_0$, i.e. the set of integers is in one to one correspondence with the set of natural numbers. An infinite set which has this property is said to be *countably infinite*.

It would appear intuitively obvious that there must be many more rational numbers than integers. On the other hand one could argue that "infinity is just infinity and the rationals cannot be more infinite than the integers". To decide between the two arguments we need to either find a one to one correspondence or prove that none exists. Simply not finding it would not be decisive. In fact, it seems quite hopeless to find it or to prove that it does not exit. However, these things have a way of working out. To explain how, we must first explain about the *Cartesian product* of two sets. It is defined by

$$A \times B = \{(a, b) | a \in A, b \in B\} \tag{1.11}$$

i.e., the set of all ordered pairs such that the first belongs to A and the second to B. Obviously, the cardinal number of the Cartesian product of two sets is the product of their cardinal numbers, i.e., $|A \times B| = |A| \times |B|$. Thus, if we consider the Cartesian product $\mathbb{N} \times \mathbb{N}$ we see that $|\mathbb{N} \times \mathbb{N}| = \aleph_0^2$.

Now consider $\mathbb{N}_1 \times \mathbb{N}_1$ (p. 19) written as the square array:

$$\mathbb{N}_1 \times \mathbb{N}_1 = \begin{matrix} (1,1), & (1,2), & (1,3), & (1,4), & \cdots \\ (2,1), & (2,2), & (2,3), & (2,4), & \cdots \\ (3,1), & (3,2), & (3,3), & (3,4), & \cdots \\ (4,1), & (4,2), & (4,3), & (4,4), & \cdots \\ \vdots & \vdots & \vdots & \vdots & \ddots \end{matrix} \} \tag{1.12}$$

We can take $(1,1)$ to correspond to $1 \in \mathbb{N}_1, (1,2)$ to $2 \in \mathbb{N}_1, (2,1)$ to $3 \in \mathbb{N}_1$, $(1,3)$ to 4, $(2,2)$ to 5, $(3,1)$ to 6, $(1,4)$ to 7, $(2,3)$ to 8, $(3,2)$ to 9 and $(4,1)$ to 10, and proceed on in this way. Clearly, to every $(p,q) \in \mathbb{N}_1 \times \mathbb{N}_1$ there will correspond some $n \in \mathbb{N}_1$. To see how this will work, notice that we are proceeding with lines which go diagonally from top right to bottom left (see Figure 1.14). The first has one element, the second two, the third three, the fourth four and so on. Thus the mth such line has m elements in it. Further (p,q) must lie in the $(p+q-1)^{st}$ line. Now the total number of terms up to and including the mth line is $m(m+1)/2$. Thus the total of all terms in the line in which (p,q) is,

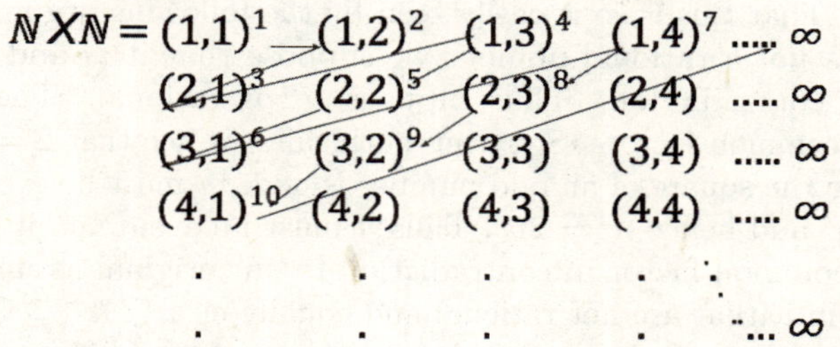

$$\begin{aligned}
\mathbb{N} \times \mathbb{N} = \quad &(1,1)^1 \quad (1,2)^2 \quad (1,3)^4 \quad (1,4)^7 \quad \ldots \infty \\
&(2,1)^3 \quad (2,2)^5 \quad (2,3)^8 \quad (2,4) \quad \ldots \infty \\
&(3,1)^6 \quad (3,2)^9 \quad (3,3) \quad (3,4) \quad \ldots \infty \\
&(4,1)^{10} \quad (4,2) \quad (4,3) \quad (4,4) \quad \ldots \infty \\
&\quad\ \ \vdots \qquad\ \ \vdots \qquad\ \vdots \qquad\ \vdots \qquad \ddots \\
&\qquad\qquad\qquad\qquad\qquad\qquad \ldots \infty
\end{aligned}$$

Figure 1.14: Starting from $(1,1)$, we move in the direction from top left to bottom right. Thus developing a procedure that counts all rationals such that none is left out.

must be $(p+q)(p+q-1)/2$. If q were 1 the natural number corresponding to (p,q) would be $(p+q)(p+q-1)/2$. For any q it will be $(q-1)$ *less* than this number. Thus, generally

$$n = \frac{1}{2}(p+q)(p+q-1) - (q-1) = \frac{1}{2}(p^2 + 2pq + q^2 - p - 3q + 2). \tag{1.13}$$

We also need to find (p,q) for a given n. For this purpose we first look for a number m such that $m(m-1)/2 < n \le m(m+1)/2$. The corresponding (p,q) must lie in the mth line. Thus $p + q = m + 1$. Also, if $n = m(m+1)/2$, $q = 1$. Thus

$$q = 1 + m(m+1)/2 - n = \frac{1}{2}(m^2 + m - 2n + 2), \tag{1.14}$$

and so

$$p = m + 1 - q = \frac{1}{2}(2n + m - m^2). \tag{1.15}$$

This completes the construction of a one to one correspondence between $\mathbb{N}_1 \times \mathbb{N}_1$ and $\mathbb{N}_1$. Hence $\aleph_0^2 = \aleph_0$ and generally $\aleph_0^n = \aleph_0$.

Now it must be clear to you how the problem with the rational numbers is resolved. Instead of writing the pairs as they are written in Eq. (1.12) we could write them as p/q. As we proceed to higher values of $p+q$ we go on eliminating those that are equivalent fractions, e.g., we retain $2/1$ but delete $4/2, 6/3, 8/4$, etc. Similarly, we retain $1/2$ but delete $2/4, 3/6, 4/8$, etc. Then again, retain $2/3$ but delete $4/6, 6/9, 8/12$, etc., and so on. The set so obtained is $\mathbb{Q}^+$. Thus $\mathbb{Q}^+$ has a one to one correspondence with a subset of $\mathbb{N}_1 \times \mathbb{N}_1$. Thus $|\mathbb{Q}^+| \le |\mathbb{N}_1|$. But $\mathbb{Q}^+ \supset \mathbb{N}_1$ and so $|\mathbb{Q}^+| \ge |\mathbb{N}_1|$. Hence $|\mathbb{Q}^+| = |\mathbb{N}_1| = \aleph_0$. Thus $\mathbb{Q} = \mathbb{Q}^+ \cup \{0\} \cup \mathbb{Q}^-$ (where $\mathbb{Q}^- = \{q| -q \in \mathbb{Q}^+\}$) has cardinality $2\aleph_0 + 1 = \aleph_0$. Thus the set of rationals is also countably infinite. This looks very much like supporting the argument that "infinity is just infinity".

We define the set of *positive algebraic numbers* as

$$\mathbb{A}^+ = \{p^q | p, q \in \mathbb{Q}^+\}, \tag{1.16}$$

where p^q stands for "p raised to the power of q" and raising to the power means repeated multiplication. (Incidentally, while $p + q = q + p$ and $p \times q = q \times p$, $p^q \neq q^p$. Also this operation is not associative, i.e., $p^{(q^r)} \neq (p^q)^r = p^{qr}$ in general.) Clearly $\mathbb{A}^+$ has a one to one correspondence with a subset of $\mathbb{N}_1 \times \mathbb{N}_1$ and hence also has cardinality $\aleph_0$. This set includes numbers like $\sqrt{2}$

which are not rational. That this is so is easily seen by the following argument.

To prove that $\sqrt{2}$ is *not* a rational number we suppose that it *is* and then prove that this supposition leads to a contradiction. Thus suppose $\sqrt{2}$ is rational. Then $\sqrt{2} = m/n$ where $m, n \in \mathbb{N}$, and have no common factors. Squaring both sides we see that $2 = m^2/n^2$ or $m^2 = 2n^2$. Hence m^2 is even. Since the square of an odd number is odd, m must be even. Thus we can write it as $2p$. Thus $m^2 = 4p^2$ and hence $n^2 = 2p^2$. Thus n must be even, i.e., it can be written as $2q$. Thus m and n have a common factor in contradiction to our original assumption. Clearly, then, $\mathbb{A}^+$ has many numbers in it that are not rational and equally clearly $\mathbb{A}^+ \supsetneq \mathbb{Q}$. Thus $|\mathbb{A}^+| \geq |\mathbb{Q}^+|$. However, since there is a one to one correspondence between $\mathbb{A}^+$ and $\mathbb{N}$, we know that $|\mathbb{A}^+| = \aleph_0$. We now obtain $\mathbb{A} = \mathbb{A}^+ \cup \{0\} \cup \mathbb{A}^-$, where $\mathbb{A}^- = \{-a | a \in \mathbb{A}^+\}$ which has cardinality $|\mathbb{A}^-| = \aleph_0$.

The question certainly arises whether there can ever be non–countable infinite sets. Let us try comparing the points in different line segments. Clearly there are infinitely many points in a line segment. It also appears clear that there would be more points in a longer line segment than in a shorter one. However, we can construct a one to one correspondence between any two line segments, see Figure 1.15. Consider the segments AB and CD. Join C to A and extend, and join

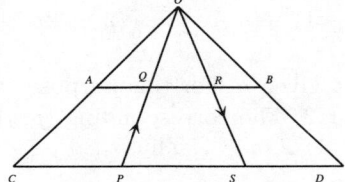

Figure 1.15: Two line segments AB and CD. CA and DB are extended so that the two intersect at O. Given $P \in CD$, PO intersects AB at Q. Given $R \in AB$, the extension of OR intersects CD at S.

D to B and extend so that the two extensions intersect at O. Now for any point $P \in CD$ we can always join it to O and intersect AB at some point Q. Similarly, given a point $R \in AB$ we can always join O to it and extend to intersect CD at S. Thus there is a one to one correspondence between the points of AB and those of CD. Thus $|AB| = |CD|$ (in the sense of cardinal numbers of the set, not in the sense of the length of the line, of course). Again 'infinity is just infinity'.

1.11 Uncountable Sets and Transfinite Numbers

Despite all the evidence presented to the contrary, so far, there are bigger and smaller infinities. For an example of a bigger infinity we go to the set of real numbers. This is a set of numbers in one to one correspondence with the points on a line. An arbitrary point O on the line is identified with 0 and the number associated with a point P, n_p, is the distance of P from O to the right. If P is to the left of O we assign a *negative* value to n_p. The line with the associated numbers is called the *real line*. (The name 'real' is unfortunate as it makes it seem as if all other number systems are unreal.) The set of real numbers, as also the real line, is generally denoted by $\mathbb{R}$. Clearly, there is an *ordering* definable on all these numbers systems. Numbers to the left, on the real line, are less than numbers to the right.

It is obvious that $\mathbb{N} \subsetneq \mathbb{Z} \subsetneq \mathbb{R}$. We now proceed to show that $\mathbb{Q} \subsetneq \mathbb{A} \subsetneq \mathbb{R}$ as well. Consider a natural number p marked on the real line. Find line segments of equal length s.t. q of them put end to end reach from O to the point given by p. These line segments have length p/q. Thus the first one, starting from O marks off p/q. This construction applies to negative numbers as well. Thus $\mathbb{Q} \subsetneq \mathbb{R}$. We already know that $\mathbb{Q} \subsetneq \mathbb{A}$. Now consider a line segment OP of unit length, see Figure 1.16. Using a compass we can construct PQ of unit length perpendicular to OP. Thus OQ

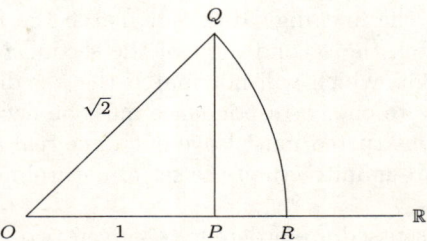

Figure 1.16: A point P on the real line represents the number 1. Using the point P as the center we can construct a perpendicular to OP and cut it at point Q such that QP is of unit length. Then OQ is of length $\sqrt{2}$. Now using O as the center and OQ as the radius we cut $\mathbb{R}$ with an arc at R. Since OR has length $\sqrt{2}$, R corresponds to the irrational number $\sqrt{2}$.

has length $\sqrt{2}$. Setting the compass point at O and its end at Q, we can cut the real line with this arc at R. Then OR has length $\sqrt{2}$ and hence R represents the number $\sqrt{2}$. Further, square roots of all natural (and rational) numbers can be similarly constructed. Cube roots etc., can also be constructed similarly using three dimensions or other arrangements. Thus $\mathbb{A} \subseteq \mathbb{R}$. In fact numbers like π can also be identified in $\mathbb{R}$ and it can be shown that they do not lie in $\mathbb{A}$. Thus $\mathbb{A} \subsetneq \mathbb{R}$. Thus $|\mathbb{R}| \geq \aleph_0$.

So far we have not found any evidence that $|\mathbb{R}| \neq \aleph_0$. To prove this fact we assume that $\mathbb{R}$ is countable. More specifically, we assume that the set of numbers greater than or equal to zero and less than one, $[0, 1)$, is countable. A given element of this set can be written as a decimal fraction. Some decimal fractions are limited, in that after some non-zero number to the right all others are zero, such as $\frac{1}{4} = 0.25$. However, others are not, such as $\frac{1}{9} = 0.111\cdots$. Nevertheless all fractions, i.e., rational numbers will lead to a recurrence of some given sequence, e.g., $\frac{1}{11} = 0.090909\cdots$, but irrational numbers will give a non–recurring infinite sequence of numbers to the right of the decimal point. Thus

$$\sqrt{2} = 1.41421356\cdots, \qquad (1.17)$$

which does not have recurrence in it. If this set is countable we can set up the following correspondence:

$$\begin{array}{ccc} \mathbb{N} & & [0, 1) \\ 1 & \longleftrightarrow & 0.a_{11}a_{12}a_{13}a_{14}\cdots \\ 2 & \longleftrightarrow & 0.a_{21}a_{22}a_{23}a_{24}\cdots \\ 3 & \longleftrightarrow & 0.a_{31}a_{32}a_{33}a_{34}\cdots \\ 4 & \longleftrightarrow & 0.a_{41}a_{42}a_{43}a_{44}\cdots \\ \vdots & & \vdots \end{array} \qquad (1.18)$$

where a_{ij} are digits between 0 and 9, $\forall i,j$. For example we could have $a_{21} = 3, a_{22} = 5, a_{23} = 0, a_{24} = 9$, etc. Again, if $\sqrt{2} - 1$ were to correspond to 7 then $a_{71} = 4$, $a_{72} = 1$, $a_{73} = 4$, $a_{74} = 2$, $a_{75} = 1$, $a_{76} = 3, a_{77} = 5$, $a_{78} = 6$ and so on. Now define a_{ij} as some number different from a_{ij}. For example $\mathsf{a}_{71} \neq 4$. We could take $\mathsf{a}_{71} = 2$, $\mathsf{a}_{72} = 0$, $\mathsf{a}_{73} = 7$, $\mathsf{a}_{74} = 5$ and so on. This notation enables us to construct the decimal fraction $0.\mathsf{a}_{11}\mathsf{a}_{22}\mathsf{a}_{33}\mathsf{a}_{44}\cdots$. This number is not in the list given in Eq. (1.18) as the first digit does not match the first digit of the first number, the second digit does not match the second digit of the second number and so on. Generally, it will have a_{nn} as the nth digit, which will not match the nth digit of the nth number. Thus our attempt to construct a one to one correspondence must be necessarily foredoomed to failure. Whatever correspondence is constructed must leave out some real numbers. Hence $|\mathbb{R}| \geq \aleph_0$, i.e., the set of real numbers is "more infinite" than the set of natural numbers. This set is said to be *continuously infinite*.

All the number systems discussed are ordered, i.e., given two numbers we can always decide which is less and which is greater. In the case of the integers they are *well ordered*, i.e., we can always tell which number comes after another and which number comes before it, i.e., given n we can find $(n+1)$ and $(n-1)$. Since there exists a one to one correspondence of $\mathbb{Q}$ with $\mathbb{N}$, rational numbers can also be well ordered. However, the ordering relation which would achieve this well ordering is not the usual ordering relationship. With the usual ordering $\mathbb{Q}$ is not well ordered. In the case of $\mathbb{R}$ it is not only not well ordered with the usual ordering but it cannot be well ordered. We could arrange it so that a next element could always be determined or that a prior element would always be identifiable, by choosing a different ordering. However, it could never be arranged that both work out, i.e., for any real number one could not find a prior *and* next element by appropriate choice of ordering.

We will not, here, go into the theory of ordered sets. However, there is a result using ordinal numbers that is relevant for our present discussion. To state the result we need to first define the power set of a given set. This is the set of all subsets of the given set. For example if the set, A, is the null set its power set $P(A)$ or $\exp(A)$ is $\{\phi\}$, i.e., a set with one element in it. If $A = \{a\}$ then $\exp(A) = \{\phi, A\}$ which has two elements in it. [Notice that the power set belongs to a higher class than the set itself. If A is of class n then $\exp(A)$ is of class $(n+1)$.] Generally, let $|A| = n$, then $|\exp(A)| = 2^n \geq n$. That this result is true for $n = 0$ or 1 has been seen. If it is true for some n it will also be true for $(n+1)$. This is seen by considering the set with n elements first and taking its power set. Now including one new element we see that we could take all the previous subsets by themselves or with the new element included. Thus the new power set will have twice as many elements as the previous power set. If the previous had 2^n elements the new one will have $2 \times 2^n = 2^{n+1}$ elements. Hence the result. This may be illustrated using a concrete example. Take $A = \{a, b, c\}$. Then $\exp(A) = \{\phi, \{a\}, \{b\}, \{c\}, \{a,b\}, \{a,c\}, \{b,c\}, A\}$ so that $|A| = 3$ and $|\exp(A)| = 8$. Now define $B = A \cup \{d\}$. Then $\exp(B) = \exp(A) \cup \{\{d\}, \{a,d\}, \{b,d\}, \{c,d\}, \{a,b,d\}, \{a,c,d\}, \{b,c,d\}, B\}$. Thus $|B| = 3 + 1 = 4$ and $|\exp(B)| = 2 \times |\exp(A)| = 16$. Generally, then, we have seen that $|\exp(A)| = 2^{|A|}$. The fact that $2^{|A|} \geq |A|$ is obviously true for finite numbers. Its validity can be demonstrated, using ordinal numbers, even for infinite numbers! This fact is crucial for our purposes.

We define $\aleph_1 = 2^{\aleph_0} \geq \aleph_0$, i.e., $|\exp(\mathbb{N})| = \aleph_1 > \aleph_0$. This procedure can be repeated indefinitely. Thus $\aleph_2 = 2^{\aleph_1} > \aleph_1, \aleph_3 = 2^{\aleph_2} > \aleph_2$, etc. Notice that the set of all subsets of $\mathbb{N}$ is the set of all integer valued, integer argument functions. Thus this set has cardinality $\aleph_1$. This sequence of numbers beyond $\aleph_0$ is called the sequence of *transfinite numbers*.

A question arises here. Is $|\mathbb{R}| = \aleph_1$ or is $|\mathbb{R}| < \aleph_1$? It seems natural to take the former but there is no *a priori* reason for doing so. The question was thoroughly investigated but no conclusion was reached [Cohen, 2008]. Since the real line is called a continuum (as opposed to the *discrete* set of integers) the hypothesis that the cardinality of $\mathbb{R}$ is $\aleph_1$ was called the *continuum hypothesis* and it was taken for granted while attempts to prove it continued. Imagine everyone's surprise when the answer to the question of the validity of the continuum hypothesis came out "yes and no". How on Earth, people felt, can mathematicians come up with such a vague and imprecise answer? However, the truth is that it is not an imprecise answer and it is not vague. It means exactly what it says.

The answer came as a corollary of *Gödel's theorem* of 1931. This theorem states that given a language constructed from a finite sufficiently large set of symbols and any finite set of axioms stated using that language there must always exist statements which cannot be derived from the given axioms. Often the validity of the statement may be independently checked, but it can not be deduced from the axioms. This theorem stands the classical view of Logic and Mathematics on its head. Everybody feels that Mathematics, at least, follows Artistotle's law of the excluded middle: "Either A is true or not A is true", where "A" is some statement. In other words any statement must be provable "true" or "false". Gödel tells us that, as far as the axioms are concerned, we can choose it to be true or false and then maintain that choice consistently! The new set of axioms must also be incomplete, in that new statements can be made which are not derivable from the axioms. It was shown that the continuum hypothesis could be chosen to be true or false! Nowadays work is being done using the negation of the continuum hypothesis. (Incidentally, there is also a lot of work being done on multiple valued logics, which are distinct from the Aristotelian 2–valued logic where every statement is either true or false.) We will assume the continuum hypothesis and proceed from there.

Clearly, with the continuum hypothesis, the set of all real valued functions with real arguments, $\mathbb{F}$, has cardinality $\aleph_2$. Since this is an important set for mathematical use, Topology will need to deal with different transfinite cardinal numbers, at least including $\aleph_0, \aleph_1$ and $\aleph_2$. There is a tendency to get confused when dealing with infinity in any case. To the extent that we get used to it, Arithmetic leads to a countable infinity, $\aleph_0$, and Geometry to the continuum, $\aleph_1$. In functional spaces, with cardinality $\aleph_2$, both arithmetic and geometric intuitions become unreliable. There we must solidly imbed ourselves in the language of sets to avoid the intuitive errors suggested by Arithmetic and Geometry.

1.12 More General Topological Transformations

Having explained the necessity for being careful in the use of geometrical arguments, let us revert to them. We had been considering two lines segments and saw that despite having different lengths we had a well defined correspondence between them. The same argument provides a correspondence between an arc of a circle and a line segment or between two wiggly lines for that matter, see Figure 1.17.

Now consider an attempt to project the circle onto the real line. The procedure is to connect the "north pole" of the circle to any point on the circle and extend to the line, e.g., NP in Figure 1.18 extended to $\mathbb{R}$ will intersect $\mathbb{R}$ at Q. The origin of $\mathbb{R}$ is fixed by joining N to S and extending, similarly, joining a point U on $\mathbb{R}$ to N will intersect the circle at V. As we come closer to N the real number obtained by this *stereographic projection* becomes larger in magnitude.

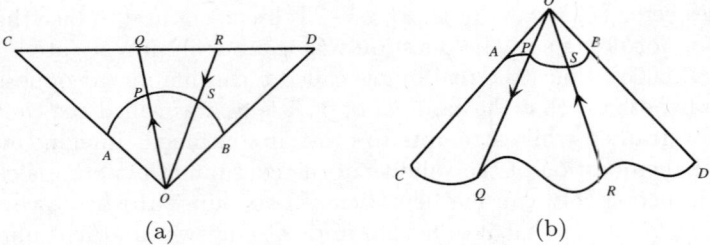

(a) (b)

Figure 1.17: (a) The mapping from arc AB to line segment CD is same as in Figure 1.15, except that AB is now curved. (b) Here both AB and CD are wiggly lines, but we still have the projection of one on to the other. These are examples of topologically equivalent spaces.

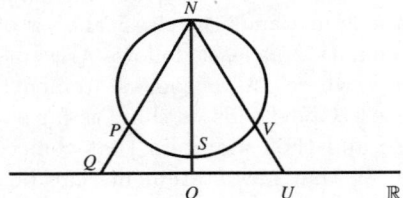

Figure 1.18: Stereographic projection of the circle, excluding the point N, onto $\mathbb{R}$ by joining points on the line to N or joining N to the points on the circle and extending.

Clearly, there is no point on $\mathbb{R}$ corresponding to N. We thus get the surprising result that the circle has more points in it than the infinite real line. However, removing the one point, N, from the circle they have the same number. This demonstrates that $|\mathbb{R}| = |[0,1)| = \aleph_1$ and since $\mathbb{R}$ can be thought of as countably infinitely many segments equivalent to $(0,1)$, $\aleph_0 \times \aleph_1 = \aleph_1$. This also provides a transformation from one space to a topologically inequivalent space.

The above example can be extended to the stereographic projection of a sphere to a plane, see Figure 1.19. Here again, it is the sphere with the north pole removed that is projected onto the plane. Notice that the ordering definable on the line, or plane, is not definable on the circle, or sphere. Further, note that while the circle is multiply connected, i.e., there are two topologically inequivalent paths between two points, the sphere is simply connected like the line and the plane. It is the torus that is multiply connected in two dimensions. One of the main aims of Topology is to study these properties and how they are related.

The ellipse, like the circle, is multiply connected and closed. The hyperbola, like the line, is simply connected and open. In fact the former two figures are topologically equivalent as are the latter two while the two pairs of figures are inequivalent. From the topological point of view we would call the former an $\mathbb{S}^1$ (a sphere of one dimension) and the latter an $\mathbb{R}^1$ (the "one" being the power of $\mathbb{R}$ in this case). Similarly, a sphere or an ellipsoid, or a more generally squashed and pulled out figure such as a cube or parallelopiped, are generically $\mathbb{S}^2$ (a sphere of two dimensions). A plane or hyperboloid of two sheets is an $\mathbb{R}^2$ (at least each of the two sheets is an $\mathbb{R}^2$, the two of them are of course two $\mathbb{R}^2$s). Here $\mathbb{R}^2 = \mathbb{R}^1 \times \mathbb{R}^1$ in the sense of the Cartesian product of sets. A cylinder, or hyperboloid of one sheet etc., is $\mathbb{R}^1 \times \mathbb{S}^1$, while a torus is an $\mathbb{S}^1 \times \mathbb{S}^1$. Notice that $\mathbb{S}^2 \neq \mathbb{S}^1 \times \mathbb{S}^1$!

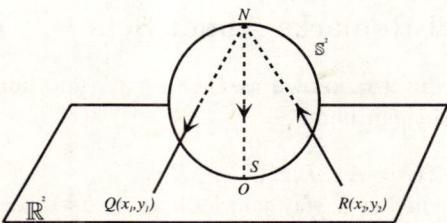

Figure 1.19: We now consider a sphere, $\mathbb{S}^2$, above a plane, $\mathbb{R}^2 = \mathbb{R} \times \mathbb{R}$. The top most point of the sphere is the "north pole", N, while the lowest point is the "south pole", S. Joining N to S and projecting gives the origin on the plane $O(0,0)$. For any point P on the sphere, with coordinates (θ_1, φ_1) we can obtain $Q(x,y) \in \mathbb{R}^2$ by joining N to P and extending. Conversely, for any $R(x_2, y_2) \in \mathbb{R}^2$ we can obtain $T(\theta_2, \varphi_2) \in \mathbb{S}^2$ by finding the point of intersection of the sphere with NR. This is the stereographic projection.

For completeness let us also define the set of *complex numbers*. These come from requiring that $\mathbb{R} \setminus \{0\}$ have a of raising to the power defined on it. Thus we get numbers like $(-1)^{\frac{1}{2}}$, etc. This is defined to be a new number called "iota" denoted by "ι" and is called an *imaginary number*. (The name is, again, most unfortunate. These numbers are no less real than the so called real numbers, which are no less sane than the rational numbers, which are no more unnatural than the natural numbers. Ironically, it is because the people naming this new number could *not* imagine it that they called it imaginary.) If we think of the real numbers placed as an east–west oriented line, the imaginary numbers $\mathbb{I} = \{ix | x \in \mathbb{R}\}$ can be depicted as a north–south oriented line running through the origin of $\mathbb{R}$. *Complex numbers* are then $\mathbb{C} = \{x + iy | x, y \in \mathbb{R}\}$, with $i^2 = -1$. Clearly, disregarding the additional structure $\mathbb{C}$ can be thought of as an ordered pair of numbers and is hence topologically equivalent to $\mathbb{R}^2$.

There are other, less common, number systems. One, for example, is the system of *quarternions* which we shall denote by $\mathbb{B}$ (for want of any other more approrprirate letter left free for use). Here we define $i^2 = j^2 = k^2 = -1$ and $ij = -ji = k$, $jk = -kj = i$, $ki = -ik = j$. Then

$$\mathbb{B} = \{p + iq + jr + ks | p, q, r, s, \in \mathbb{R}\}. \tag{1.19}$$

This system was investigated by William Rowan Hamilton and can be used to deal with a Minkowski geometry. There had been great hopes that it would play a fundamental role in uniting the two major physical theories: the Quantum Theory; and General Relativity, but the hope remains unfulfilled. After this there are octonians, $\mathbb{O}$, but none further [Conway and Smith, 2003].

It is necessary, here, to consider the problem of ordering on $\mathbb{C}$ or $\mathbb{R}^2$. Given two numbers in $\mathbb{R}^2$, $p = (p_1, p_2)$ and $q = (q_1, q_2)$ it may not be possible to decide which is greater even when they are unequal. The problem arises if $p_1 > q_1$ and $p_2 < q_2$. Thus, according to one of the variables $p > q$ and according to the other variable $p < q$. Hence this set cannot be meaningfully regarded as ordered. However, we can always find an $r > p, q$ by choosing $r = (r_1, r_2)$ so that $r_1 > p_1 > q_1$ and $r_2 > q_2 > p_2$. Similarly, we can choose $s = (s_1, s_2)$ such that $s_1 < q_1 < p_1$ and $s_2 < p_2 < q_2$ so that $s < p, q$. A set in which an ordering is not necessarily defined between all elements but for which, given any two elements, we can always find elements less than both and greater than both, is said to be *partially ordered*.

1.13 Some Additional Remarks About Sets

There are some useful rules about sets known as *DeMorgan's laws* formulated by the Augustus De Morgan. We shall briefly state them here:

1. $A \cap (B \cap C) = (A \cap B) \cap C = A \cap B \cap C$
   which simply says that whichever way you look at it the three–way intersection consists of those elements belonging to all three sets (and hence intersections are *associative*);

2. $A \cup (B \cup C) = (A \cup B) \cup C = A \cup B \cup C$,
   i.e., union is associative;

3. $A \cap B = B \cap A$,
   since both sides consist of elements belonging to both sets (and hence intersection is *commutative*);

4. $A \cup B = B \cup A$,
   i.e., union is commutative;

5. $A \cup (B \cap C) = (A \cup B) \cap (A \cup C)$,
   as the left side consists of those elements which either lie in A or in both B and C, while the right side has elements that either lie in A or B and also either lie in A or C and hence lie in A or in B and C (thus union is *distributive* over intersection);

6. $A \cap (B \cup C) = (A \cap B) \cup (A \cap C)$,
   due to a similar argument, i.e. intersection is distributive over union;

7. $(A \cap B)' = A' \cup B'$,
   as the left side consists of those elements which do not lie in both A and B while the right side consists of those elements which either do not lie in A or do not lie in B;

8. $(A \cup B)' = A' \cap B'$,
   as the left side consists of those elements which do not lie in either A or B and the right side of those elements which do not lie in A and do not lie in B.

(The latter two statements are what are generally referred to as De Morgan's laws.) These laws can be easily followed using Venn diagrams, introduced by John Venn. The Venn diagram is a convenient pictorial representation of sets. The universal set is denoted by a rectangle and any subset of it by a circle inside it, see Figure 1.20. The complement of the set is then the region *outside* the circle. The intersection of two sets is the region of overlap of the two circles. The union of the two sets is the region covered by both circles. If the two sets have no region in common they are *disjoint* sets and their intersection is the null set. The various laws stated above are now seen very simply using Venn diagrams.

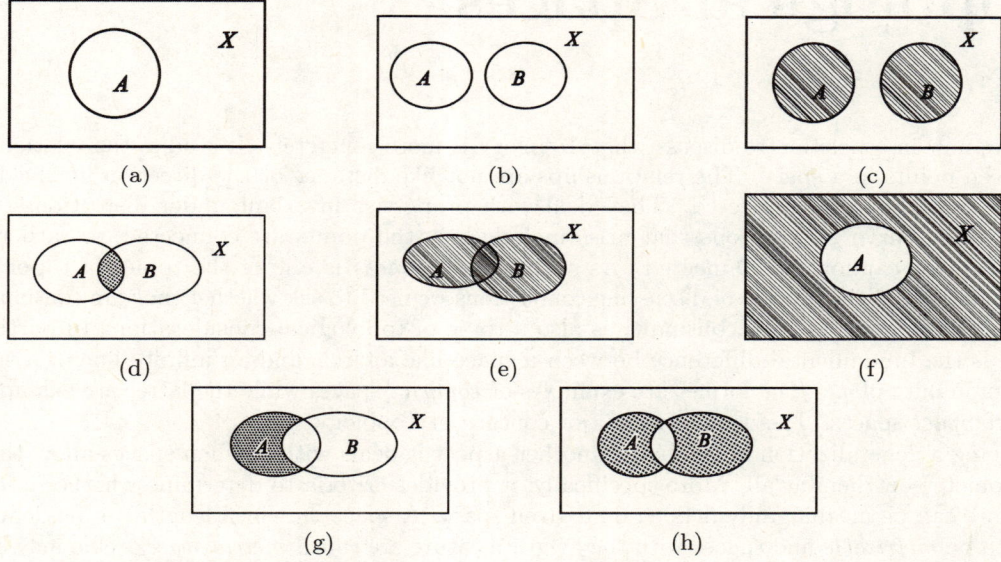

Figure 1.20: Venn diagrams depict the universal set X as a box and subsets as circles inside the box. (a) $A \subsetneq X$; (b) $A, B \subsetneq X$, $A \cap B = \phi$; (c) $A, B \subsetneq X$, $A \cap B = \phi$ and $A \cup B$ is shown by the shaded region; (d) $A, B \subsetneq X$, $A \cap B \neq \phi$ is depicted by the shaded region; (e) $A, B \subsetneq X$, $A \cup B$ is denoted by the shaded region and $A \cap B \neq \phi$; (f) $A, \subsetneq X$, A' is shown as the shaded region; (g) $A, B \subseteq X$, $A \cap B \neq \phi, A \setminus B$ is shown as the shaded region; (h) $(A \cup B) \setminus (A \cap B)$ is shown as the shaded region.

Chapter 2

Topological Spaces

We are now in a position to discuss what Topology is more concretely. It studies the relationships between points in a space. The relationships are not like distance or the direction involved, but rather how they are connected. They deal with properties invariant under distortions of the "connecting curve". Questions that arise include how the points are connected, or rather how they can (or cannot) be connected. As such *connectedness* is one of the topics of topological investigation. In the process of discussing connections we need to ask whether they are continuously connected or not. As such continuity is also a topic of topological consideration. In particular there is the fundamental difference between a space like a circle and an infinite line or a sphere and an infinite plane. The former are examples of *compact* spaces while the latter are examples of *non-compact* spaces. This is another major concern of Topology.

Being a generalization of geometry another aspect it deals with is when spaces allow the use of geometry on them at all. More specifically, it provides criteria to determine whether a length measure can be meaningfully defined on a given space. A space on which length can be defined is said to be *metrizable* and spaces with the length measure are called *metric spaces*. One may define a magnitude in a space *without* it providing a measure of length. This magnitude is generally called a *norm* and the space with this norm defined on it is called a *normed space*. There may be spaces for which there is no metric or norm definable. However, provided some basic criteria are met we can still study it topologically. This general study of spaces is the subject of this chapter.

2.1 The Definition of a Topological Space

To be able to talk of continuity we need to take limits. Thus, in the part of the space under consideration, we need that we should be able "to go either side" of a given point. In an intuitive sense this requirement is provided by what is called an *open interval*, but not by a *closed interval*. For example, if we consider the set of numbers between 0 and 1 and excluding 0 and 1, denoted by $(0,1) = \{x | 0 \lneq x \lneq 1\}$, we can always find numbers on either side of any $x \in (0,1)$. This is an *open set* . On the other hand, in the case of the *closed set* $[0,1] = \{x | 0 \leq x \leq 1\}$, we cannot find any $x \in [0,1]$ such that $x < 0$ and we cannot find any $x > 1$. At the base of Topology is this concept of an "open set".

In a general setting the meaning to be assigned to an "open set" is arbitrary. However, there are some requirements that must carry over from the open interval. An obvious requirement is that the union of two open sets be open. If this did not hold we could find that on a part of the space we could talk of continuity meaningfully, on another part we could also do so, but putting the two together we would be unable to do so. Thus we would be unable to extend our analysis to the entire space. As such, our definition of "open sets" *must* be consistent with the requirement that the union of the sets also be open.

As a simple example, if $A = (0, 2)$ and $B = (1, 3)$ then $C = A \cup B = (0, 3)$. Now A and B are open intervals and so is C, see Figure 2.1. Instead consider a non-obvious example. Take $X = \{a, b, c\}$. Suppose we decide that only $\{a\}$ and $\{b\}$ will be called open subsets of A and all other subsets are closed. Now $\{a\} \cup \{b\} = \{a, b\}$ will *not* be open, in contradiction to our original requirement. This will not be an appropriate generalization of the open interval we were in search of. For consistency, then, we require that the *union of open sets is open*.

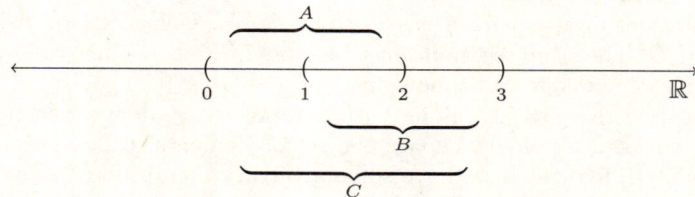

Figure 2.1: The union of $A = (0, 2)$ and $B = (1, 3)$ is $C = (0, 3)$ which is an open interval.

Similarly, we must require that the *intersection of open sets is open*. In the previous example of $A = (0, 2)$, $B = (1, 3)$ we will have $D = A \cap B = (1, 2)$. Again, in the example $X = \{a, b, c\}$, if we chose $\{a, b\}$ and $\{b, c\}$ to be the only open sets, apart from X itself, then $\{a, b\} \cap \{b, c\} = \{b\}$ would contradict the requirement. Thus the requirement is again non-trivial. The above requirements were stated pairwise. We could easily extend the requirements to many sets. Thus, if A, B, C are open $A \cup (B \cup C) = A \cup B \cup C$ must also be open and $A \cap (B \cap C) = A \cap B \cap C$ must also be open. In fact, if $A_i \subseteq X (i = 1, \cdots, n)$ are open sets then $\sqcup_{i=1}^n A_i$ and $\bigcap_{i=1}^n$ are also open, where $\sqcup_{i=1}^n A_i = A_1 \cup A_2 \cup \cdots \cup A_n$ and $\bigcap_{i=1}^n A_i = A_1 \cap A_2 \cap \cdots \cap A_n$. This extension is obvious, as we could reduce the whole union (or intersection) to pairwise disjoint unions (or intersections). The question is whether the above extension would hold for an infinite number of subsets, i.e., $n \to \infty$. The indices may be denoted, for dealing with infinitely many subsets, by $i \in I$, where I is the *index set*, i.e., the set of all indices. We could have $I = \mathbb{N}$, $I = \mathbb{N}_e$, $I = \{1, \cdots, n\}$, $I = \mathbb{R}$, etc.

There is no reason why there should be any problem developing with the union of more and more open sets, even going to infinitely many. After all no new "end points" of the sets can come in this way. However, if we take an infinite sequence of open subsets, they may turn out to have only one point in common. For example, take $\mathbb{R}$ to be the universal set and define $A_n = \{x | -\frac{1}{n} < x < \frac{1}{n}\}$. Clearly, the set is $A_n = (-\frac{1}{n}, \frac{1}{n})$, which is open, see Figure 2.2. However, if we now take $\bigcap_{n \in \mathbb{N}} A_n$ we see that any $x \neq 0$ will lie *outside* this intersection for some n. Thus, $\bigcap_{n \in \mathbb{N}} A_n = \{0\}$, which is a singleton set. Clearly this is not open in our usual definition. Thus, we should only expect the requirement to hold for a *finite* intersection of subsets.

We are now in a position to define our set of all open subsets of the universal set. This set is called the *topology*, denoted by τ, for our space X. We allow τ to be *an arbitrary subset of* $\mathbb{P}(X)$ whose elements satisfy the following requirements: (a) *the union of all elements of τ belongs to τ*; (b) *a finite intersection of all elements of τ belongs to τ*. In general, there can be disjoint subsets

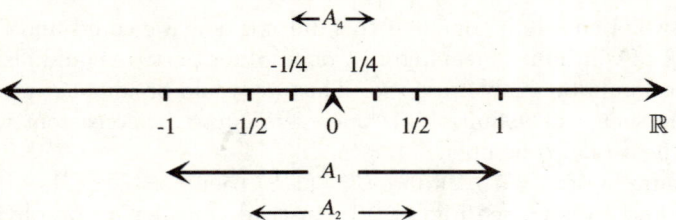

Figure 2.2: In the real line, $\mathbb{R}$, we define open subsets as $A_1 = (-1, 1)$, $A_2 = (-\frac{1}{2}, \frac{1}{2})$, $A_3 = (-\frac{1}{3}, \frac{1}{3})$ (not shown here), $A_4 = (-\frac{1}{4}, \frac{1}{4})$ and so on. Clearly $\bigcap_{n \in \mathbb{N}} A_n = \{0\}$.

of X in τ. By requirement (b) their intersection, ϕ, must also belong to τ. Hence the empty set ϕ is open. The complement of an open set is said to be *closed*. Consider two disjoint closed proper subsets of X, C and D. Their intersection must be closed. Thus ϕ must be closed. Thus $\phi' = X$ must be open. Hence $\phi, X \in \tau$ for all topologies.

A *topological space* is a space X, along with a topology, τ, defined on it. It is denoted by the pair (X, τ). In words, a topological space is a set of elements (called points), with some well defined sense associated, for which we have some arbitrary definition of what we mean by open and closed sets. The definition must satisfy certain consistency requirements. These requirements force us to regard both the empty and the universal sets as open. A moment's reflection shows us that ϕ and X must also be closed since $\phi' = X$ and $X' = \phi$. Thus ϕ, X are both open and closed! This fact seems paradoxical at first, since "closed" seems to be the opposite of "open". However, it will be remembered that the null set was defined for consistency and cannot be expected to satisfy usual intuitive requirements based on non-empty sets. Similarly, there is no meaning assignable to "openness" or "closedness" for the whole space. As an example, consider the economy of a country. It can be open or closed to outside trade. However, the economy of the world as a whole cannot be said to be "open to outside trade" or "closed to it", as there is no concept of "outside trade" for it. There is nothing outside the universal set. As such the concept of "open" or "closed" does not really apply to it. We are, therefore, free to choose to call it "open" or "closed" or both, or neither. For consistency of our definitions we choose to call the null set and the universal set both open and closed. We could now, talk of two topological spaces (X_1, τ_1) and (X_2, τ_2) and ask when (and in what way) they differ or are the same. Before doing so we will need to develop the language of Topology further.

2.2 Examples of Topologies

To clarify the concept of a topology we consider some specific examples of topologies. There is, of course, the *usual topology* defined on the set of real numbers by identifying open sets with open intervals. This was, of course, the concept we wanted to generalize to more unusual spaces. We start with examples whose sole merit is to be simple, and so (hopefully) easy to follow. We proceed on to various topologies that are more commonly referred to.

Example 1. State all possible topologies for $X = \{a\}$.
Solution: The power set of X is $\mathbb{P}(X) = \{\phi, X\}$. Since these elements belong to every topology there is a unique topology for this set, namely $\tau = \mathbb{P}(X)$.

Example 2. Find all possible topologies for $X = \{a, b\}$.
Solution: $\mathbb{P}(X) = \{\phi, \{a\}, \{b\}, X\}$. We have the topologies

$$\tau_1 = \{\phi, X\}, \tau_2 = \{\phi, \{a\}, X\}, \tau_3 = \{\phi, \{b\}, X\}, \tau_4 = \mathbb{P}\{X\}. \tag{2.1}$$

In general the first and last topologies can always be defined. The former, τ_1, is called the *indiscrete topology* and is denoted by τ_i, while the latter, τ_4 here, is called the *discrete topology* and is denoted by τ_d. In both cases all open sets are also closed. In the latter case, in fact, all sets are both open and closed. A space with a discrete topology is called a *discrete space* and with an indiscrete topology an *indiscrete space*. It is obvious that the indiscrete topology is the smallest and the discrete topology is the largest possible topology for a given space. Note that in Example 1, $\tau_d = \tau_i$.

Example 3. Give all possible topologies for $X = \{a, b, c\}$.
Solution: In the previous examples the number of distinct topologies was $2^{|X|}$. This was true because all subsets of $\mathbb{P}(X)$, containing $\{\phi\}$ and $\{X\}$ were topologies. That is not the case here. There were cases of subsets of $\mathbb{P}(X)$ that are not topologies. For instance, $\{\phi, \{a\}, \{b\}, X\}$ is not a topology as the union of the supposed open sets $\{a\}$ and $\{b\}$ is not an open set. For acceptable topologies, apart from the two cases that must always be there, we have $\{\phi, \{a\}, X\}$ and the corresponding topologies containing $\{b\}$ and $\{c\}$ which we denote collectively by τ_{10} and three more like $\{\phi, \{a, b\}, X\}$ which we denote by τ_{01}. Then we have nine τ_{11} like $\{\phi, \{a\}, \{a, b\}, X\}$ or $\{\phi, \{a\}, \{b, c\}, X\}$. Then there are three τ_{21} like $\{\phi, \{a\}, \{b\}, \{a, b\}, X\}$. Then there are three τ_{12} like $\{\phi, \{a\}, \{a, b\}, \{a, c\}, X\}$ and six τ_{22} like $\{\phi, \{a\}, \{b\}, \{a, b\}, \{a, c\}, X\}$. Thus there are 29 distinct topologies. [How many subsets will not be topologies?]

Question: If $|X| = n$, how many distinct topologies can be made?

Example 4. For a non-empty X, taking any arbitrary $A \subsetneq X$, a topology other than the two always definable can be constructed: $\tau = \{\phi, A, A', X\}$. In this case all open sets are also closed.

Example 5. Again with $A \subsetneq X$ we can define the A−*inclusion topology*, $\tau_1 = \{\phi, B \supseteq A\}$ see Figure 2.3. This is the topology of all *supersets* of A. We also have the A−*exclusion topology* $\tau_2 = \{\phi, B \subsetneq X, X | B \cap A = \phi\}$, see Figure 2.4. In Example 3, the τ_{21} given is the $\{c\}$−exclusion topology. Again, with A as above $\tau_3 = \{\phi, B \subseteq A, X | A \subsetneq X\}$ we have the topology of all subsets of A. It is easy to verify that this is the A'−exclusion topology.

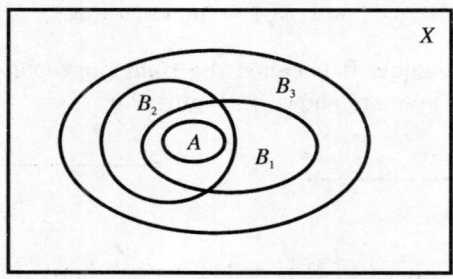

Figure 2.3: The A–inclusion topology in Venn diagrams. All B in which A lies will be called "open sets" and no others will.

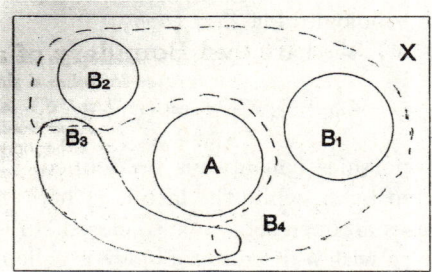

Figure 2.4: The A–exclusion topology in Venn diagrams. All B which do not touch A will be called "open sets" and no others will.

Example 6. With $X = \mathbb{R}$ define

$$\tau_c = \{\phi, A \subsetneq X \,|\, |A'| = n \in \mathbb{N}\}, \tag{2.2}$$

i.e., the complement of an open set is defined to have finite cardinal number. This is called the *co-finite topology*. To verify that this is a topology note that $X \in \tau$ as $X' = \phi$ and $|\phi| = 0 \in \mathbb{N}$. Further, define an index set, I, for the subsets belonging to the topology, then we require that $\bigcup_{\alpha \in I} A_\alpha \in \tau$. Now $\left(\bigcup_{\alpha \in I} A_\alpha\right)' = \bigcap_{\alpha \in I} A'_\alpha$ by De Morgan's laws. Thus, obviously $\left|\bigcap_{\alpha \in I} A'_\alpha\right| \leq n_\alpha$, where $|A'_\alpha| = n_\alpha$, implies that the requirement holds. Similarly we require that $\bigcap_{\alpha \in I} A_\alpha \in \tau$. Again using De Morgan's laws $\left(\bigcap_{\alpha \in I} A_\alpha\right)' = \bigcup_{\alpha=1}^{m} A'_\alpha$. Thus we have $\left|\left(\bigcap_{\alpha \in I} A_\alpha\right)'\right| \leq \sum_{\alpha=1}^{m} n_\alpha$ (where m is some finite number). This demonstrates that the other requirement is also met. If, instead, we define

$$\tau_2 = \{\phi, A \subseteq X \,|\, |A'| = \aleph_0\}, \tag{2.3}$$

We have the co-countable topology. (Prove that τ *is* in fact a topology.) For a finite set X, the co-finite topology is simply the discrete topology. Similarly, for a countable set X, the co-countable topology is the discrete topology.

Example 7. Take $X = \mathbb{R}$,

$$\tau = \{\phi, A, \mathbb{R} | A = (a, \infty), a \in \mathbb{R}\}. \tag{2.4}$$

It is obvious that this is a topology. It is called the *right ray topology*. Instead, if $A = (-\infty, a)$, we get the *left ray topology*. These are shown in Figure 2.5.

Figure 2.5: Only one proper subset of $\mathbb{R}$ is called "open" here. (a) The "right ray topology", where A is the ray to the right of a; (b) the "left ray topology", where A is the ray to the left of a.

2.3 The Interior, Closure and Boundary of a Set

In general a subset, A, in a topological space (X, τ) need not be either open or closed. For example the interval $[0, 1)$ is neither open nor closed. Another example is of a disc $D = \{(x, y) | x^2 + y^2 \leq 1$ and the equality holds only for $x \notin \mathbb{Q}\}$, in the real plane $X = \mathbb{R} \times \mathbb{R} = \mathbb{R}^2$. As such there will be infinitely many points of the boundary included and infinitely many excluded. Both types are mixed together in a way that cannot be visualized easily. We need to generalize this concept to spaces with lower and higher cardinality where visualization may not be easy. Further, we could construct the *largest open set* contained in the given set, $(0, 1)$, and the *smallest closed set* containing it, $[0, 1]$. We need to generalize all these concepts. We formally define an interior point of a set A as $x \in A$ s.t. $\exists\, U \subseteq A$ and $U \in \tau$ s.t. $x \in U \subseteq A$. We define the interior of A, by

$$int(A) = \{x | x \in U \subseteq A, U \in \tau\}, \qquad (2.5)$$

i.e., the set of all interior points of A. As an example, taking $\mathbb{R}$ with open sets given by open intervals and $A = [0, 1)$, we see that $int(A) = (0, 1)$. Since $A \setminus int(A) = \{0\}$ we need only verify that $\{0\} \notin int(A)$. This is true as any $U \subseteq A$ which contains 0 cannot be open. As another example consider the topology constructed for $X = \{a, b, c\}, \tau = \{\phi, \{a\}, \{b\}, \{a, b\}, X\}$. Take $A = \{a, b\}$. Here $A \in \tau$. Thus we can see that $a \in \{a\} \subseteq A, b \in \{b\} \subseteq A, \{a, b\} \subseteq A$. Thus the interior of A is A. Now consider $A = \{a, c\}$. Here $a \in \{a\} \subseteq A$ and $\{a\} \in \tau$ but $c \in X \in \tau$ only. Thus $c \notin int(A)$. Hence, in this case $int(A) = \{a\}$. Similarly, if $A = \{b, c\}$, $int(A) = \{b\}$. The above examples lead us to expect the following result.

Theorem 2.1. Given a topological space (X, τ) and $A \subseteq X$

(i) $int(A) \in \tau$

(ii) $A = int(A) \Leftrightarrow A \in \tau$.

Proof. The first statement is obvious. Nevertheless we can proceed to prove it by *reductio ad absurdum*, to demonstrate the way such theorems are proved. Suppose $int(A) \notin \tau$. Then $\exists\, x \in int(A)$ s.t. $x \notin U \subseteq A$ where $U \in \tau$. But this contradicts the definition of $int(A)$. Hence $int(A) \in \tau$. This proves (i). To prove (ii) we note that the forward implication is obvious by (i). Further, if $A \in \tau$ we can choose $U = A \in \tau$. Thus $A = int(A)$. This proves the reverse implication and hence completes the proof of (ii). We notice that this shows that $int(A)$ is the largest open subset of A in that $\nexists\, B \in \tau$ s.t. $int(A) \subsetneq B \subsetneq A$. $\square$

Let us now proceed to the closed set containing A. As mentioned earlier, a set is said to be *closed* if its complement is open. (The natural belief that proper subsets cannot be both open and closed is based on our intuition derived from intervals. However, there is no reason why it should not be possible in a more general context.) We can, thus, consider $[int(A)]' = X \setminus int(A)$, which must be closed. However, $int(A') = int(X \setminus A)$ is open by the above theorem. Their difference is the set of "end points" around A, see Figure 2.6. It is called the *boundary* (or *frontier*) of A and denoted by ∂A. Thus

$$\partial A = [int(A)]' \setminus int(A'). \qquad (2.6)$$

We define the *closure* of A, by
$$\bar{A} = A \cup \partial A. \tag{2.7}$$
Clearly, if $A \in \tau$, $A \cap \partial A = \phi$ but if $A \notin \tau$ then $A \cap \partial A \neq \phi$. $\bar{A}$ is the smallest closed set containing A. You are encouraged to try to prove this result rigorously. Also, obviously
$$\partial A = \bar{A} \smallsetminus int(A), \tag{2.8}$$
∂A being closed in the sense of the topology τ.

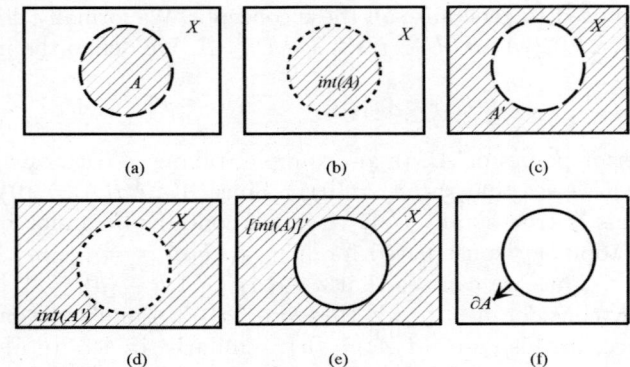

Figure 2.6: Using Venn diagrams we can construct the boundary of a set. (a) A set A is "partially open" and "partially closed", the hard line being the "end points" included in the set and the broken one being excluded; (b) the interior of A is the open set in which the end points are excluded; (c) $A' = X \smallsetminus A$ shown in the same way; (d) $int(A')$ shown; (e) $[int(A)]'$ shown; (f) the difference between (e) and (d) is the boundary of A, $\partial A = [int(A)]' \smallsetminus int(A')$.

To clarify these concepts let us consider some examples. We start with the example $A = [0, 1)$ mentioned above for which $int(A) = (0, 1)$. Now proceeding with the definitions
$$\begin{aligned}[int(A)]' &= \{x | x \in \mathbb{R} \text{ and } x \leq 0 \text{ or } x \geq 1\} \\ &= (-\infty, 0] \cup [1, \infty).\end{aligned} \tag{2.9}$$
We also have $A' = (-\infty, 0) \cup [1, \infty)$. Thus
$$int(A') = (-\infty, 0) \cup (1, \infty). \tag{2.10}$$
which gives the boundary of A as
$$\partial A = [int(A)]' \smallsetminus int(A') = \{0, 1\}. \tag{2.11}$$
Thus we also have the closure of A
$$\bar{A} = A \cup \partial A = [0, 1) \cup \{0, 1\} = [0, 1]. \tag{2.12}$$
Notice that $\bar{A}$ is also equal to $int(A) \cup \partial A$ and $\bar{A} \smallsetminus A = \{1\}$.

As another example consider the disc, D, defined above (at the start of this section), see Figure 2.7. Clearly, we have
$$int(D) = \{(x,y)|x^2 + y^2 < 1\}. \tag{2.13}$$
Thus we get
$$[int(D)]' = \{(x,y)|x^2 + y^2 \geq 1\}. \tag{2.14}$$
$$D' = \{(x,y)|x^2 + y^2 \geq 1 \text{ with the equality holding for } x \notin \mathbb{Q}\}, \tag{2.15}$$
$$int(D') = \{(x,y)|x^2 + y^2 > 1\}, \tag{2.16}$$
which yields the boundary of the disc as the circle
$$S' = \partial D = [int(D)]' \smallsetminus int(D') = \{(x,y)|x^2 + y^2 = 1\}. \tag{2.17}$$
The closure of the disc is then
$$\bar{D} = D \cup \partial D = int(D) \cup \partial D = \{(x,y)|x^2 + y^2 \leq 1\}. \tag{2.18}$$
Notice that here
$$\bar{D} \smallsetminus D = \{(x,y)|x^2 + y^2 = 1 \text{ and } x \notin \mathbb{Q}\}. \tag{2.19}$$
As an exercise it would be useful to construct the interior, boundary and closure of an arbitrary non-open subset for some choice of topology in Example 3 of Section 2.2. For example, take $A = \{a, c\}$ with the topology $\tau = \{\phi, \{a\}, \{a, b\}, X\}$. The result, in this case, is $int(A) = \{a\}, \partial A = \{b, c\}, \bar{A} = X$.

We shall return to the concepts of closure and boundaries later.

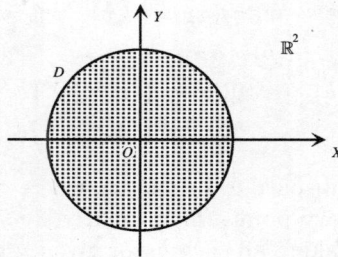

Figure 2.7: The disc $D \subsetneq \mathbb{R}^2$. Here only boundary points with a rational value of x lie in D and all the points with an irrational value of x on the unit circle are excluded. Clearly, $D \notin \tau$ here.

2.4 Neighborhoods and Neighborhood Systems

The concept of "neighborhood" as the surroundings in ordinary usage needs to be made more precise. In ordinary usage it is implicit that the neighborhood does not extend too far but it is not clear what "too far" signifies. As such we have to dispense with this connotation in Topology. All we require is that no region should be included in the neighborhood which is blocked off from the place whose neighborhood is being identified. Also, if some place is at the border of a region, part of its neighborhood must lie *outside* that region. Formally, then, a set A can be a *neighborhood* (written *nbd* for short) of a point x if $\exists\, U \in \tau$ s.t. $x \in U \subseteq A$. In other words A is a *nbd* of x if x is an interior point of A (see Figure 2.8). A simple example from real intervals is that $(-1, 1)$ is a *nbd* of 0. So, for that matter, is $(-2, 2)$ or $(-\frac{1}{2}, 3)$, and so on. An arbitrary *nbd* of x will generally be denoted by $\eta(x)$ and specific *nbds* by $\eta_1(x), \eta_2(x)$, etc. The set of all *nbds* of x, $\mathfrak{N}(x)$, is called the *nbd system* of x. Let us consider some examples to clarify this concept.

Figure 2.8: A region A is a neighborhood of a point x if we can find an open set U inside A as shown in (a). If x lies on ∂A then clearly an open set U containing x cannot lie inside A as shown in (b).

Example 1. For the topological space

$$X = \{a, b, c\},\ \tau = \{\phi, \{a\}, \{a, b\}, \{a, c\}, X\} \tag{2.20}$$

find the *nbds* of a, b and c and give their *nbd* systems.
Solution: Clearly X is a *nbd* of every point. Also we have

$$\eta_1(a) = \{a\}, \qquad \eta_2(a) = \{a, b\}, \qquad \eta_3(a) = \{a, c\}, \tag{2.21}$$
$$\therefore \mathfrak{N}(a) = \{\{a\}, \{a, b\}, \{a, c\}, X\}. \tag{2.22}$$
$$\eta_1(b) = \{a, b\}. \quad \therefore \mathfrak{N}(b) = \{\{a, b\}, X\}. \tag{2.23}$$
$$\eta_1(c) = \{a, c\}. \quad \therefore \mathfrak{N}(c) = \{\{a, c\}, X\}. \tag{2.24}$$

Example 2. Take the topological space to be $(\mathbb{R}, \tau_c)$, where τ_c is the co-finite topology defined in Eq. (2.2) and $a \in \mathbb{R}$ any arbitrary point. Then clearly $\mathbb{R} \setminus \{b_1, \cdots, b_n\}$ is a *nbd* of a for any choice of $b_i \neq a$ for any finite n. Taking all such b_i for given n gives $\mathfrak{N}_n(a)$.

Example 3. In the topological space $(\mathbb{R}, \tau_u)$ where τ_u stands for the usual topology defined in Section 2.2, we have four types of *nbds* of a point x, namely

$$\left.\begin{array}{l}\text{(i)}\quad \eta_1(x) = (a, b),\\ \text{(ii)}\quad \eta_2(x) = [a, b),\\ \text{(iii)}\quad \eta_3(x) = (a, b],\\ \text{(iv)}\quad \eta_4(x) = [a, b].\end{array}\right\} \text{s.t. } a, b \in \mathbb{R} \text{ and } a < x < b. \tag{2.25}$$

The first is, itself, an open set. Thus we have $x \in \eta_1(x) \subseteq \eta_1(x)$ and $\eta_1(x) \in \tau_u$. The second is again a *nbd* as $\eta_1(x) \subseteq \eta_2(x)$, and so is the third. Finally the fourth is automatically a *nbd* as $\eta_1(x) \subseteq \eta_2(x), \eta_3(x) \subseteq \eta_4(x)$. Thus a *nbd* need not be an open set itself, but must contain an open set containing the given point. Notice that $\eta_2(x)$ is *not* a *nbd* of x for $x = a$. To see this note that any $a \in U \in \tau_u$ is of the form $U = (a - \delta, a + \epsilon)$ with $\delta, \epsilon > 0$ and $\delta, \epsilon \in \mathbb{R}$ which is impossible for $U \subseteq [a, b)$. If $A \subseteq B$ then $A \smallsetminus B = \phi$ by definition. Here $U \smallsetminus [a, b) = (a - \delta, a) \neq \phi$. Hence $[a, b)$ is not a *nbd* of a. This type of reasoning can be applied more generally.

2.5 Isolated Points

In analytical geometry planar curves are defined by functional relations between two variables, $\varphi(x, y) = 0$. These are generally taken to be single valued functions, $y = f(x)$, but may give branches, segments and isolated points. Since Topology is primarily motivated by trying to generalize geometry to spaces that are vastly different from the plane (where geometry was born) we will be interested in defining the topological extensions of these concepts. The segments and branches will come later. Here we are interested in generalizing the concept of an isolated point. For this purpose consider the curve defined by

$$\varphi(x, y) \equiv x^3 - x^2 y + x^2 + xy^2 - y^3 + y^2 = 0. \tag{2.26}$$

This may be simplified to the equation

$$(x^2 + y^2)(x - y + 1) = 0. \tag{2.27}$$

One would be tempted to regard the straight line $y = x + 1$ as the unique solution of this equation. However, the solution set also includes the point $(0, 0)$ which makes the expression in the first bracket in Eq. (2.27), zero. Thus the curve, C, given by Eq. (2.26) consists of the straight line, L, and a point O (see Figure 2.9). The point O does not lie on L and is not, therefore, connected to the rest of the curve. It is called an isolated point.

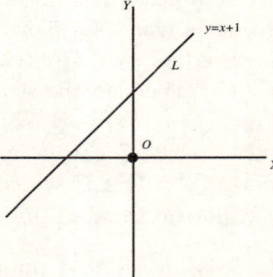

Figure 2.9: The "curve" given by Eq. (2.26) is the line L and the point O, i.e., $C = L \cup O$. Clearly O is isolated from the rest of the curve. It is called an isolated point in geometry.

To generalize the above concept let us first analyze what we had there. The topological space was $(\mathbb{R}^2, \tau_u)$. Now $C \subsetneq \mathbb{R}^2$ and $C = L \cup O$. There existed regions in $\mathbb{R}^2$ which could totally surround O without intersecting L. Thus $\exists\, U \subsetneq \mathbb{R}^2$ s.t. $U \in \tau_u$ and $C \cap U = O$. In fact we also had $A \supseteq U$ s.t. $C \cap A = O$. Of course, not every A would satisfy this requirement. For example $A = \{(x, y) | x^2 + y^2 < 1\}$ would not. All we require is that $\exists$ *some nbd* A of the point s.t. $C \cap A = O$. We can choose, for example, $A = \{(x, y) | x^2 + y^2 < \frac{1}{2}\}$ as the *nbd*.

Generally, for any topological space (X, τ) and a set $C \subseteq X$ if $\exists\, x_0 \in C$ s.t. $\exists\, \eta \in \mathfrak{N}(x_0)$ s.t. $C \cap \eta = \{x_0\}$ then x_0 is said to be an *isolated point* of C. Let us look at some examples of isolated points, and sets that have no isolated points, in this more general setting.

Example 1. Consider the topological space (X, τ_d), where τ_d is the discrete topology defined in Example 2 of Section 2.2. Given any $x \in X$ we know that $\{x\} \in \tau_d$. Thus $\{x\}$ is a *nbd* of x and $\{x\} \cap X = \{x\}$. Thus every element of X is an isolated point of X. The entire space is composed of isolated points. This is why it is called a discrete topology. Instead, consider (X, τ_i) where τ_i is the indiscrete topology. Here $\forall x \in X, \eta(x) = X$ is the only *nbd* of the element. Thus any set $C \ni x$ will have $C \cap \eta(x) = C \cap C = C \neq \{x\}$, unless $C = \{x\}$. Thus apart from singleton sets no set can contain an isolated point. Obviously a singleton set consists of an isolated point (by definition).

Example 2. In $(\mathbb{R}, \tau_u), \mathbb{Z}$ consists of isolated points. To see this consider $n \in \mathbb{Z}$. We can construct $A = (n - \delta, n + \epsilon)$ with any $0 < \delta, \epsilon < 1$. As $A \in \tau_u$ and $n \in A$ we can take A to be a *nbd* of n. Now $\mathbb{Z} \cap A = \{n\}$ and so n is an isolated point of $\mathbb{Z}$, regardless of the choice of n. In other words n is the only integer in the real interval $(n - \delta, n + \epsilon)$, see Figure 2.10.

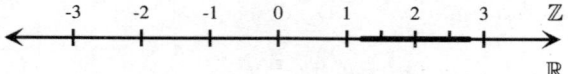

Figure 2.10: To illustrate the point we take $n = 2$ and $\epsilon = 0.7 = \delta$, so that we have the open interval $(1.3, 2.7) \in \mathbb{R}$ which contains the number 2. Now $\mathbb{Z}$ is the sequence of numbers above the real line, which correspond to some points in $\mathbb{R}$. Clearly the intersection of $\mathbb{Z}$ and $\mathbb{R}$ consists of $\mathbb{Z}$, and intersection of $\mathbb{Z}$ and $(1.3, 2.7)$, consists of $\{2\}$.

Example 3. Whereas $\mathbb{Q}$, like $\mathbb{Z}$, is countable, $\mathbb{Q}$ has no isolated points in the topological space $(\mathbb{R}, \tau_u)$. To prove this claim let us consider any $x \in \mathbb{Q}$. Now, consider any $a, b \in \mathbb{R}$ s.t. $a < x < b$, i.e., (a, b) is a *nbd* of x. Now we can always find $a_1 \in \mathbb{Q}$ s.t. $a < a_1 < x$ as we could round off a to some decimal fraction smaller than x. (This is called the Archimedean property of real numbers.) Similarly, $\exists\, b_1 \in \mathbb{Q}$ s.t. $x < b_1 < b$. Thus $\exists\, \{a_1, b_1\} \in \mathbb{Q}$, corresponding to any $(a, b) \subsetneq \mathbb{R}$, which will then give $(a_1, b_1) \cap \mathbb{Q} \neq \{x\}$. Hence no $x \in \mathbb{Q}$ can be an isolated point of $\mathbb{Q}$. Similarly, had we chosen the set of irrational numbers, $\mathbb{Q}' = \mathbb{R} \smallsetminus \mathbb{Q}$ we could have proved that it also has no isolated points. Even more obviously $\mathbb{R}$ has no isolated points in the usual topology.

Example 4. While every $n \in \mathbb{Z} \subsetneq \mathbb{R}$ is an isolated point of $\mathbb{Z}$ in τ_u, this is not true in the co-finite topology. Remember that in this topology a set is open if its complement is a finite set. Consider any $n \in \mathbb{Z}$. Now we can construct *nbds* of n. For example, choose $m \neq n$ and $m \in \mathbb{Z}$. More generally, take any $x \in \mathbb{R}$ s.t. $x \neq n$. Then $\mathbb{R} \smallsetminus \{x\}$ is a *nbd* of n and its intersection with $\mathbb{Z}$ is at $\{m | m \neq x \text{ and } m \in \mathbb{Z}\} \neq \{x\}$. Now consider the complement of $\{x, y\}$, $\mathbb{R} \smallsetminus \{x, y\}$, which intersects $\mathbb{Z}$ at $\{m \in \mathbb{Z} | m \neq x, m \neq y\} \neq \{n\}$. In fact, if we consider the complement of any finite set it will intersect $\mathbb{Z}$ at infinitely many points. Hence $\mathbb{Z}$ has no isolated points in the co-finite topology. However, in the *co-countable topology* we can choose a *nbd*

$$\eta(n) = \mathbb{R} \smallsetminus (\mathbb{Z} \smallsetminus \{n\}) \tag{2.28}$$

of n which intersects $\mathbb{Z}$ only at n, making n an isolated point of $\mathbb{Z}$.

2.6 Some Simple Topological Theorems

There are various statements that can be made about topological properties of spaces. Some statements are very profound. Some have very far reaching consequences. Some are very difficult to prove. Traditionally, all such statements used to be called "theorems" and direct consequences of these theorems were called "corollaries". Occasionally a result was needed to prove a conjectured theorem to be valid and was called a "lemma". In more recent times there have been lemmas of more fundamental importance than most theorems and many theorems too trivial to be generally known. The boundary between the "lemma", the "theorem" and the "corollary" has grown very indistinct. Also, the original difference between the "lemma" and the "theorem" was not in the statement itself but in the intention of the person proving it. It was any how a matter of the judgement (and opinion) of the person stating it as to which statement had great significance, which was only a minor consequence of the result and which was only needed to prove the main result. Without apology we will call various statements "theorem" and others "corollaries", following our judgement or that of others in the field. The statements will occasionally be obvious and not either profound or significant, but will be called theorems for want of any thing better to call them.

Theorem 2.2. A topological space (X, τ) is discrete iff every point, $x \in X$, is an isolated point.

Proof. As we have already seen the former statement implies the latter. Thus the former is a *necessary condition* for the latter to be true. We need to prove the reverse implication, i.e., that it is a *sufficient condition* as well. For this purpose assume that every $x \in X$ is isolated. Then $\forall x \in X, \exists U \in \tau$ s.t. U is a *nbd* of x and $U \cap X = \{x\}$, which can happen only if $U = \{x\}$. Hence $\{x\} \in \tau \ \forall \ x \in X$, which means that the topology is discrete. $\square$

The following characterization of open sets often proves very useful.

Theorem 2.3. For a given $A \subseteq X$ in a topological space (X, τ), $A \in \tau$ iff $\forall x \in A, A$ is a *nbd* of x.

Proof. Since A consists only of interior points (to be an open set), A is a *nbd* of x if $A \in \tau$. Conversely if A is a *nbd* of x, $\forall x \in A, \exists U_x \in \tau$ s.t. $x \in U_x \subseteq A$. Hence $\{x | x \in A\} \subseteq \cup \{U_x | x \in A\} \subseteq A$, i.e., $A \subseteq \cup \{U_x | x \in A\} \subseteq A$. Thus $A = \cup \{U_x | x \in A\}$. Since A is a union of open sets it is open. $\square$

Theorem 2.4. For $\eta_1, \eta_2 \in \mathfrak{N}(x) \Longrightarrow$

(i) $\eta_1 \cup \eta_2 \in \mathfrak{N}(x)$;

(ii) $\eta_1 \cap \eta_2 \in \mathfrak{N}(x)$;

(iii) $(A \subseteq B$ and $A \in \mathfrak{N}(x)) \Longrightarrow B \in \mathfrak{N}(x)$.

Proof. To prove (i) and (ii) consider $\eta_1, \eta_2 \in \mathfrak{N}(x)$. Then $\exists U_1, U_2 \in \tau$ s.t. $x \in U_1 \subseteq \eta_1$ and $x \in U_2 \subseteq \eta_2$. Thus $x \in U_1 \cup U_2 \subseteq \eta_1 \cup \eta_2$ and $x \in U_1 \cap U_2 \subseteq \eta_1 \cap \eta_2$. Since $U_1 \cup U_2$ and $U_1 \cap U_2$ are the union and intersection, respectively, of open sets they are open. Hence $\eta_1 \cup \eta_2$ and $\eta_1 \cap \eta_2$ are *nbds* of x, which proves (i) and (ii). To prove (iii) assume that the left side of the implication is valid. Then $\exists A_1 \in \tau$ s.t. $x \in A_1 \subseteq A$ and $A \subseteq B$. Hence $x \in A_1 \subseteq B$ and B is a *nbd* of x, which proves (iii). $\square$

Theorem 2.5. For a given $A \subseteq X$ in a topological space (X, τ)

$$int(A) = \cup\{U | U \in \tau \text{ and } U \subseteq A\}. \tag{2.29}$$

Proof. If $x \in int(A)$ then $\exists U \in \tau$ s.t. $x \in U \subseteq A$. Hence $x \in \cup\{U | U \in \tau, U \subseteq A\}$. Conversely, if $x \in \cup\{U | U \in \tau, U \subseteq A\}$ then $x \in U \subseteq A$ for some $U \in \tau$. Thus if x belongs to the set on the left side of Eq. (2.29) it also lies in the set on the right side and vice verse. Therefore the two sets are equal. Clearly, $int(A)$ is the largest open set in A. □

The above theorems make it easy to work out the interior of a given set. For example in $(\mathbb{R}, \tau_u)$ it is clear that $int(\mathbb{Z}) = \phi$ as there is no open subset of $\mathbb{Z}$ in this topology. Similarly, from Theorem 2.3, since for any $x \in \mathbb{Q}$ every nbd of $x, \eta(x)$, must intersect $\mathbb{R}$, no subset of $\mathbb{Q}$ is a *nbd* of x, i.e., if $A \subseteq \mathbb{Q}$ then $A \ni x$ is *not* a *nbd* of x. Hence $A \notin \tau$. Thus $int(\mathbb{Q}) = \phi$. Similarly $int(\mathbb{Q}') = \phi$. As is obvious we could use the theorems to work out that $int([a, b]) = (a, b), int((a, b)) = (a, b)$ and in fact $int(\mathbb{R}) = \mathbb{R}$. More generally, of course, in any topological space, $int(X) = X$ and since the open set is its own interior $int(int(A)) = int(A)$.

Now consider $(\mathbb{R}, \tau_c)$ as the topological space. Clearly, for any $a \in \mathbb{R}, (\mathbb{R} \smallsetminus \{a\}) \in \tau_c$. Thus $int(\mathbb{R} \smallsetminus \{a\}) = \mathbb{R} \smallsetminus \{a\}$. In fact $int(\mathbb{R} \smallsetminus \{b_1, b_2, \ldots, b_n\}) = \mathbb{R} \smallsetminus \{b_i\}$ where $i = 1, \cdots, n$ for some $n \in \mathbb{N}$. However, $int(\mathbb{R} \smallsetminus \mathbb{Z}) = \phi$ as $(\mathbb{R} \smallsetminus \mathbb{Z}) \notin \tau_c$. In the *co-countable topology* $(\mathbb{R} \smallsetminus \mathbb{Z}) \in \tau$ and hence $int(\mathbb{R} \smallsetminus \mathbb{Z}) = \mathbb{R} \smallsetminus \mathbb{Z}$. Clearly, $\mathbb{Z} \notin \tau_c$ and in fact $\forall A \subseteq \mathbb{Z}, A \notin \tau_c$. Thus $int(\mathbb{Z}) = \phi$. It is worth proving for yourself that $int([a, b])$ in this topology is ϕ. In fact, generally, $int(A) = A$ if $A \in \tau$ and $int(A) = \phi$ if $A \notin \tau$.

Theorem 2.6. (i) $A \subseteq B \implies int(A) \subseteq int(B)$;

(ii) $\forall A_\alpha \subseteq X, \alpha \in I, \bigcup_{\alpha \in I}(int A_\alpha) \subseteq int(\bigcup_{\alpha \in I} A_\alpha)$,
$\bigcap_{\alpha \in I}(int A_\alpha) \subseteq int(\bigcap_{\alpha \in I} A_\alpha)$. (If I is finite the equality holds.)

Proof. (i) is obvious as $int(A)$ is the largest open set contained in the set. Since $int(A)$ is open and $int(A) \subseteq A \subseteq B$, it is obvious that $int(A)$ can at most be $int(B)$. For (ii) note that $int(A_\alpha)$ could be ϕ for each α but the union of A_α may be open. For example $int(\mathbb{Q}) = int(\mathbb{Q}') = \phi$ in τ_u, but $(\mathbb{Q} \cup \mathbb{Q}') = int(\mathbb{R}) = \mathbb{R}$. The rigorous proof is left as an exercise for you. □

2.7 Topology in Terms of Closed Sets

So far we have developed all topological concepts in terms of open sets, merely defining closed sets and the closure of sets. This concept is *dual* to the concept of open sets and topology could equally well have been developed starting with the definition of closed sets only. This duality is brought out in the following theorem.

Theorem 2.7. For any topological space (X, τ):

(i) ϕ and X are closed sets;

(ii) the intersection of closed sets is closed; and

(iii) a finite union of closed sets is closed.

Proof. Since $\phi, X \in \tau$ and $\phi' = X, X' = \phi$. Hence X and ϕ are closed by definition. This proves (i). Now we have

$$\bigcup_{a \in I} A_\alpha \in \tau \quad \text{if all} \quad A_\alpha \in \tau. \tag{2.30}$$

Using De Morgan's law $\left(\bigcup_{\alpha \in I} A_\alpha\right)' = \bigcap_{\alpha \in I} A'_\alpha$ is closed and, by definition A'_α are all closed. This proves (ii). Again

$$\bigcap_{\alpha=1}^{n} A_\alpha \in \tau \quad \text{if all} \quad A_\alpha \in \tau. \tag{2.31}$$

Again, De Morgan's law gives $\left(\bigcap_{\alpha=1}^{n} A_\alpha\right)' = \bigcup_{\alpha=1}^{n} A'_\alpha$ is closed. This proves (iii). Notice that in the dualization here the *unions* had to be finitely many while the *intersections* could be infinitely many. □

Example 1. In (X, τ_d) all $A \subseteq X$ are open as well as closed. Hence $\bar{A} = A = int(A)$ and $\partial A = \phi$. In (X, τ_i) no proper subset of X is either open or closed. Hence $int(A) = \phi, \bar{A} = X \, \forall A \subseteq X$. Thus $\partial A = X$.

Example 2. In $(\mathbb{R}, \tau_u), \mathbb{Q}, \mathbb{Q}', [a, b), (a, b]$ are neither open nor closed. However for $\mathbb{Z}$ (or any subset of $\mathbb{Z}$), $[a, b]$ are closed. To verify these latter statements note that

$$\mathbb{Z}' = \cdots \cup (-2, -1) \cup (-1, 0) \cup (0, 1) \cup (1, 2) \cdots, \tag{2.32}$$

which is open. The same applies for any subset. Similarly, $[a, b]' = (-\infty, a) \cup (b, \infty)$ is open. Thus $int(\mathbb{Z}) = \phi, \partial \mathbb{Z} = \mathbb{Z}$.

Example 3. In $(\mathbb{R}, \tau_c)$ every finite subset is closed as

$$\mathbb{R} \smallsetminus \{b_1, \cdots, b_n\} = \{b_1, \cdots, b_n\}' \in \tau. \tag{2.33}$$

Thus $\partial \{b_1, \cdots, b_n\} = \{b_1, \cdots, b_n\}$.

It is worthwhile to emphasize that the dual concepts of open and closed are *not* mutually exclusive, unlike the usual usage where something open is, by definition, not closed. Here we have seen repeated examples of sets that are both open and closed in a given topology and of sets that are neither open nor closed in some topology. In the usual topology the concepts are mutually exclusive for proper subsets, but even there ϕ and X are both open and closed.

Finally, it should be mentioned that though closed sets *could* have been used equally well to characterize topologies, the convention adopted has been to use open sets.

2.8 Limit Points, the Derived Set and Perfect Sets

The concept of the boundary of a set can be arrived at from a different approach than that taken in Section 2.3. Though that approach may seem simpler, in being easier to grasp, it is by no means easy to use to calculate the boundary of a set. To be able to define the boundary in the new way we need a series of definitions which we now proceed to provide.

A point $x \in X$ is said to be a *limit* (or an *accumulation*) point of a subset $A \subseteq X$, if every nbd, η of x contains some point of A other than x. Thus $(\eta \smallsetminus \{x\}) \cap A \neq \phi$. It is not necessary that $x \in A$. Of course, if $x \in A \neq \{x\}$, x will be a limit point of A unless $\{x\} \in \tau$. If $\{x\}$ is an open set then $\exists \ \eta = \{x\}$ which is a *nbd* of x and hence $\eta \smallsetminus \{x\} = \phi$. However, even if $x \notin A$, it is possible that every *nbd* of x must partly lie in A. For example, if we consider $A = (0,1) \subsetneq \mathbb{R}$ with τ_u as the topology $0 \notin A$ but every $\eta(0)$ will intersect A.

The set of all limit points of a set A is called the *derived set* of A and is denoted by $\mathfrak{D}(A)$. Clearly, there can be points of A that do not lie in $\mathfrak{D}(A)$, provided the singleton sets containing those points are open. Also, there can be points in $\mathfrak{D}(A)$ that are not in A. Thus it is *not* true that $A \subseteq \mathfrak{D}(A)$ or that $A \supseteq \mathfrak{D}(A)$ in general. Let us consider some examples of derived sets.

Example 1. For a discrete space, (X, τ_d), $\mathfrak{D}(X) = \phi$ as every point is an isolated point and hence $\exists \ \eta$ s.t. $\eta \smallsetminus \{x\} = \phi$. Similarly, for an indiscrete space, (X, τ_i), $\mathfrak{D}(A) = X \ \forall \ A \subseteq X$.

Example 2. In $(\mathbb{R}, \tau_u)$, $\mathfrak{D}(\mathbb{Z}) = \phi$ as $\mathbb{Z}$ consists of only isolated points in this topology. However $\mathfrak{D}(\mathbb{Q}) = \mathbb{R}$, as near every rational number there is a real number. Again $\mathfrak{D}((a,b)) = [a,b]$ while $\mathfrak{D}([a,b]) = [a,b]$. Take $A = \{\frac{1}{n}|n \in \mathbb{N}\}$. Then 0 is a limit point of A. In fact it is the only limit point of A which does not lie in A. Consider the curve C given by Eq. (2.26). O is *not* a limit point of C while the whole of L is. Thus $\mathfrak{D}(C) = L = C \smallsetminus O$.

Example 3. For $(\mathbb{R}, \tau_c)$, $\mathfrak{D}(\mathbb{Z}) = \mathbb{R}, \mathfrak{D}(\mathbb{N}) = \mathbb{R}, \mathfrak{D}(\mathbb{Q}) = \mathbb{R}$. It would be useful to go through the procedure of verifying these statements. The hint, for the first two, is contained in the discussion of these sets in the context of isolated points in Section 2.5.

It is clear that the closure of A could also be computed from the derived set by $\bar{A} = A \cup \mathfrak{D}(A)$ as the boundary points have been included in the limit points and though the isolated points were thrown away in $\mathfrak{D}(A)$ they are included in A. Further, if $A = \mathfrak{D}(A)$ then A must be a closed set as $\bar{A} = A$ in that case. Such a set is called a *perfect set*. Not every closed set is a perfect set as we can have $A \smallsetminus \mathfrak{D}(A) \neq \phi$. A closed set which is not perfect could have isolated points.

For a discrete space $\bar{A} = A \ \forall \ A \subseteq X$ while for an indiscrete space $\bar{A} = X \ \forall \ A \subseteq X$. In $(\mathbb{R}, \tau_u)$, $\bar{\mathbb{Z}} = \mathbb{Z}$ while $\bar{\mathbb{Q}} = \mathbb{R}$ and $\bar{\mathbb{Q}'} = \mathbb{R}$. In fact if $|A| \in \mathbb{N}$ then $\forall A \subseteq \mathbb{R}, \bar{A} = A$. In $(\mathbb{R}, \tau_c), \bar{\mathbb{Z}} = \mathbb{R}, \bar{\mathbb{Q}} = \mathbb{R}$ as is obvious from the previous example. Also $\overline{(a,b)} = \overline{[a,b]} = \mathbb{R}$ in $(\mathbb{R}, \tau_c)$.

Example 4. Consider the topological space (X, τ) with topology $\tau = \{\phi, A, A', X\}$. Clearly, every element of the topology is both open and closed. Hence we see that $\bar{A} = A$ and $\bar{A'} = A'$. Thus $\mathfrak{D}(A) = A$ and both A and A' are perfect sets. Also $int(A) = A$. In this case, as in the discrete topology, we have $int(A) = A = \mathfrak{D}(A) = \bar{A}$ and $\partial A = \phi \ \forall \ A \subseteq X$. However, for $B \subseteq X$ s.t. $B \neq A$ or $B \neq A', \bar{B} = X, int(B) = \phi, \partial B = X$.

We can state the dual theorem to Theorem 2.5 for closed, instead of open sets. Here we obviously get the following statement.

Theorem 2.8. For a given $A \subseteq X$ in a topological space (X, τ)

$$\bar{A} = \cap \{F | F' \in \tau, F \supseteq A\}. \tag{2.34}$$

Proof. The simplest way to see this result is to take the complement of Eq. (2.29), use De Morgan's laws and the definition of closed sets. $\square$

From the above theorem it is clear that $\bar{A}$ is the smallest closed *superset* (set containing the given set) of A. As we saw with the interiors, if $A \subseteq B$ then $\bar{A} \subseteq \bar{B}$. The fact that inclusion, as a relation between sets, carries through for interiors or closures is expressed by saying that these operations are *monotonic*. Again, as with $int(A)$ we have $\bar{\bar{A}} = \bar{A}$ as $\bar{A}$ is the smallest closed set containing A so $\bar{\bar{A}}$ is the smallest closed set containing the closed set $\bar{A}$. But $\bar{A}$ *is* the smallest such closed set. Hence the assertion.

Example 5. Take $X = \{a,b,c\}, \tau = \{\phi, \{a\}, \{a,b\}, X\}$ and $A = \{a\}$. Then b is a limit point of A as there are two *nbds* of b, $\eta_1(b) = \{a,b\}, \eta_2(b) = X$. Clearly, $(\eta_i(b) \smallsetminus \{b\}) \cap A = \{a\} \neq \phi$. Again c is a limit point of A as $\exists! \ \eta(c) = X$ and $(\eta(c) \smallsetminus \{c\}) \cap A = \{a\} \neq \phi$. Thus $\mathfrak{D}(A) = X$ and hence $\bar{A} = X$ and $\partial A = \{b,c\}$ as seen before, at the end of Section 2.3. You should try working out these quantities with other topologies with the same X.

Note that dual to the statement that only a finite number of intersections of open sets is guaranteed to be open, only a finite number of unions of closed sets is guaranteed to be closed. For instance, in Example 3 we saw that the closure of $\mathbb{Z}$ (or $\mathbb{N}$ for that matter) is $\mathbb{R}$, while for $A_n = \{-n, \cdots, 0, \cdots, n\}$ is A_n, i.e., $\bar{A}_n = A_n$ but $\bar{\mathbb{Z}} = \mathbb{R}$, in the co-finite topology. Thus $\bigcup_{n=1}^{\infty} \bar{A}_n = \mathbb{Z}$. However, $\overline{\bigcup_{n=1}^{\infty} A_n} = \bar{\mathbb{Z}} = \mathbb{R}$. Thus $\overline{\bigcup_{n=1}^{\infty} A_n} \neq \bigcup_{n=1}^{\infty} \bar{A}_n$. Here, incidentally, $\partial \mathbb{Z} = \mathbb{R}$.

It is worth while to point out that we could re-write Eq. (2.34), using DeMorgan's law, as

$$(\bar{A})' = int(A'), \tag{2.35}$$

by the definition of $int(A)$. Conversely,

$$[int(A)]' = \overline{A'}. \tag{2.36}$$

The details of the proof are left as an exercise for you.

Using Eq. (2.35) with Eq. (2.6) we see that

$$\partial A = \overline{A'} \smallsetminus int(A'). \tag{2.37}$$

From Theorem 2.5 and the remarks succeeding Eq. (2.7) we see that

$$\partial A \cap int(A) = \phi. \tag{2.38}$$

Thus, from Eq. (2.8) it is obvious that

$$\partial A = \bar{A} \cap \overline{A'}. \tag{2.39}$$

Since $\bar{\phi} = \phi = int(\phi)$ and $\bar{X} = X = int(X), \partial \phi = \bar{\phi} \smallsetminus int(\phi) = \phi$ and $\partial X = \bar{X} \smallsetminus int(X) = \phi$. In fact generally, if $\bar{A} = A = int(A)$ then $\partial A = \phi$. Conversely, if $\partial A = \phi, \bar{A} \smallsetminus int(A) = \phi$. Thus $\bar{A}$ has no element not contained in $int(A)$, i.e., $\bar{A} \subseteq int(A)$. But $int(A) \subseteq A \subseteq \bar{A}$. Hence $\bar{A} = A = int(A)$. Thus $\partial A = \phi$ iff A is both open and closed. Now $int(A') \cap int(A) = \phi$ is obvious from Figure 2.6. Thus

$$X = int(A) \cup \partial A \cup int(A') \tag{2.40}$$

implies that

$$X \smallsetminus \partial A = int(A) \cup int(A'), \tag{2.41}$$

and hence $(\partial A)'$ is open (being a union of open sets), so ∂A is always closed. From here we see that $\overline{\partial A} = \partial A$ and hence $\partial(\partial A) = \overline{\partial A} \smallsetminus \partial A = \phi$. Thus unlike $\overline{\overline{A}} = \overline{A}$ and $int(int(A)) = int(A)$, the boundary operation does *not* have $\partial(\partial A) = \partial A$.

2.9 Dense Sets and Separable Spaces

As repeatedly stressed, our whole purpose is to generalize geometry to deal with different spaces, even to the extent of having different cardinalities. However, it is as well to bear in mind that we have a special interest in a space on which the usual concepts of Riemannian and Lobachevskian geometry apply. In this section we set up the basic language using which we can *topologically* specify the types of spaces on which geometry can be used. Of course, geometry here includes the requirement that we can use ordinary calculus. (We will, later, be referring to some examples of *extra*-ordinary calculus from a topological point of view.)

To be able to give one of the properties of an "acceptable space" for geometrical purposes we need to first define a relationship between a subset and the universal set of a given topological space. $A \subseteq X$ is said to be *dense* in X if $\overline{A} = X$. Thus, obviously, X *is* dense in X. Again, if $X = [0,1]$ then $A = (0,1)$ is dense in X. Similarly, if $X = \{(x,y)|x^2 + y^2 \leq 1, x, y \in \mathbb{R}\}$ then $D = \{(x,y)|x^2 + y^2 < 1, x, y \in \mathbb{R}\}$ is dense in X. In the indiscrete topology $\tau_i = (\phi, X)$, clearly every A is dense in X. This is easy to see. Since $int(A) = \phi$ and $int(A') = \phi$, so $\partial A = [int(A)]' \smallsetminus [int(A')] = \phi' \smallsetminus \phi = X \smallsetminus \phi = X$. Thus $\overline{A} = int(A) \cup \partial A = A \cup \partial A = X$. In some sense these are trivial examples. We have simply chosen sets for which the "outer edge" is the boundary, or which are (by definition) the space itself. Only the last example could be regarded as more interesting, as there it was not *immediately* obvious. The boundary of the set was *not* the "outer edge" but included the set itself. This gives a hint of the type of cases which are of special interest as they lead to the type of definitions we require for our "acceptable spaces".

Example 1. In (X, τ_d), other than X itself, no subset is dense in X. This is obvious as every subset of X is both open and closed. Hence every subset is its own closure, i.e., $\forall A \subsetneq X, A = \overline{A} \neq X$.

Example 2. For $(\mathbb{R}, \tau_c)$, no finite set is dense in $\mathbb{R}$ and every infinite subset is dense in $\mathbb{R}$. This is clear as $\forall A \subsetneq \mathbb{R}$ s.t. $|A| \in \mathbb{N}, \overline{A} = A$, while $\forall A \subsetneq \mathbb{R}$ s.t. $|A| = \aleph_0$ or $\aleph_1, \overline{A} = \mathbb{R}$. In particular, $\mathbb{N}, \mathbb{Z}$ and $(0,1)$ are dense in $\mathbb{R}$ with this topology.

Example 3. In $(\mathbb{R}, \tau_u), \overline{\mathbb{N}} = \mathbb{N} \neq \mathbb{R}$ and in fact $\overline{\mathbb{Z}} = \mathbb{Z} \neq \mathbb{R}$. Thus even these infinite sets are not dense in $\mathbb{R}$, while finite sets are obviously not dense in $\mathbb{R}$. However, both $\mathbb{Q}$ and $\mathbb{Q}'$ *are* dense in $\mathbb{R}$ as $\overline{\mathbb{Q}} = \mathbb{R} = \overline{\mathbb{Q}'}$. *This* is the promised non-trivial example. The fact that the rationals are dense in $\mathbb{R}$ makes it possible for us to construct sequences (numerically) with which we can take limits to calculate derivatives to any arbitrary accuracy. Had we been forced to use only non-computable numbers to take limits for the purposes of using calculus, there could not have been any definition of differentials or differential coefficients by actual computation. For completeness we remark, also, that $\mathbb{A}$ and $\mathbb{A}'$ (the set of algebraic numbers and its complement defined in Section 1.10) are also dense in $\mathbb{R}$. This is obvious since $\mathbb{Q} \subsetneq \mathbb{A} \subsetneq \mathbb{R}$ and hence $\overline{\mathbb{Q}} \subset \overline{\mathbb{A}} \subset \overline{\mathbb{R}} = \mathbb{R}$. Since $\overline{\mathbb{Q}} = \mathbb{R}$ it is obvious that $\overline{\mathbb{A}} = \mathbb{R}$.

We have a *topological* specification of a space on which calculus can be used. It is useful to have alternative ways of characterizing sets with a given property. In particular, it may not always be so easy to see whether $\bar{A} = X$ or not. We can try using the following theorem in that case.

Theorem 2.9. In a topological space (X, τ), a subset A is dense in X iff every non-empty open set U, contains a point of A.

Proof. The first implication is seen directly as there are no points not in the *nbd* system of A, since $\mathfrak{N}(A) = \{X\}$. (By the *nbd* of a set, $\eta(A)$ is meant a set which is the *nbd* of all its elements. The *nbd* system, $\mathfrak{N}(A)$, is the set of all *nbds* of the set A. Here it has only one element as the *smallest* closed set containing A is X.) Conversely, if $\forall a \in A \; \exists \, U \in \tau$ s.t. $U \in \mathfrak{N}(A)$ then $U \subseteq \bar{A}$. Thus $\bigcup_{U \in \tau} U \subseteq \bar{A}$. But $\bigcup_{U \in \tau} U = X$. Hence $X \subseteq \bar{A} \subseteq X$. Thus $\bar{A} = X$. □

The significance of the theorem may be further clarified by re-stating it in negative form. If there is no way to choose an open set that does not contain a point of A, then A is dense in X. Here are various statements that are often useful when dealing with dense sets (proofs of which are left as an exercise for you):

(1) $\forall B \supseteq A, (A$ is dense in $X) \implies (B$ is dense in $X)$;

(2) $(A$ is dense in $X)$ and $(B$ is open and dense in $X) \implies (A \cap B$ is dense in $X)$;

(3) In $(X, \tau), (A_i \in \tau (i = 1, \cdots, n)$ are dense in $X) \implies (\bigcap_{i=1}^{n} A_i$ is dense in $X)$.

To see the significance of requiring B to be open in (2), it would be enough to see a counterexample to the statement *without* this requirement. More or less the first one that may be thought of provides the desired counterexample. Since $\mathbb{Q}$ and $\mathbb{Q}'$ are neither open nor closed in $(\mathbb{R}, \tau_u)$ we take them. $\mathbb{Q} \cup \mathbb{Q}' = \mathbb{R}$, but while $\mathbb{Q}, \mathbb{Q}'$ are dense in $\mathbb{R}, \mathbb{Q} \cap \mathbb{Q}' = \phi$ is not. It is obvious that (3) is a direct generalization of (2). The significance of limiting the index set to be finite should be pondered by you, following the same line of reasoning as given above for (2).

Theorem 2.10. In a topological space (X, τ), if A is dense in X and $U \in \tau$ then $\overline{U} = \overline{A \cap U}$.

Proof. Now $\overline{U} \supseteq \overline{A \cap U}$ is obvious so we only need to prove that $\overline{U} \subseteq \overline{A \cap U}$. For this purpose consider $x \in \overline{U}$ (though x may lie on ∂U and not in U) and an arbitrary $\eta(x) \in \tau$ s.t. $x \in \eta(x)$. As is apparent from Figure 2.8, $U \cap \eta(x) \neq \phi$ and by definition $U \cap \eta(x) \in \tau$. Thus, by Theorem 2.9, $(U \cap \eta(x)) \cap A \neq \phi$. As intersections are associative and commutative $\eta(x) \cap (A \cap U) \neq \phi$. Hence $x \in \overline{A \cap U}$. Since this is true for any arbitrary $x \in X, (x \in \overline{U}) \implies (x \in \overline{A \cap U})$. Thus $\overline{U} \subseteq \overline{A \cap U}$ and hence $\overline{U} = \overline{A \cap U}$. (Notice that it was already obvious that $\overline{U} = \overline{A} \cap \overline{U} = X \cap \overline{U}$ but it is not so obvious that $\overline{A \cap U} = \overline{A} \cap \overline{U}$. In fact this is not generally true. Consider $A = (-1, 0)$ and $U = (0, 1)$ in $(\mathbb{R}, \tau_u)$. Then $A \cap U = \phi$ and so $\overline{A \cap U} = \phi$ but $\overline{A} \cap \overline{U} = \{0\}$. It is the fact that A is dense in X that guarantees the equality.) □

We are now in a position to state the requirement for a space on which we can perform calculus operations for geometrical purposes using the usual geometrical concepts. Loosely speaking, one would say that the space should not have *too many* points nor *too few* points. A space is said to be *separable* if there exists a countably infinite subset of X which is dense in it, i.e., if $\exists A \subseteq X$ s.t. $\bar{A} = X$ and $|A| = \aleph_0$. Obviously, every countably infinite space is separable as X is always dense in X. However, if $|X| \geq \aleph_1$ then (X, τ_d) is non-separable while (X, τ_i) is separable. $(\mathbb{R}, \tau_u)$ is obviously separable as $\overline{\mathbb{Q}} = \mathbb{R}$ and $|\mathbb{Q}| = \aleph_0$. Again $(\mathbb{R}, \tau_c)$ is separable as $\overline{\mathbb{N}} = \mathbb{R}$ in that topology.

This is *not* an adequate requirement for making the space acceptable but it is necessary. Geometry requires separability but is not guaranteed by it.

2.10 Topological Bases

In cases where there are infinitely many elements of the topology it is obviously not possible to enumerate them all. As such it may not be possible to determine whether two specifications of a topology on the same set, X, refer to the same topological space or two different topological spaces. Even with finite sets the number of possible topologies may be so large as to make it impossible to verify by enumeration. Let us first look into the question of how many topologies can be defined on a finite set.

If $|X| = n$ then the number of possible subsets, i.e., the number of elements in the discrete topology is 2^n since the topology is a subset of the power set. The number of possible topologies must, then, be less than or equal to the cardinal number of the power set of the power set of X, i.e., $2^{(2^n)}$. However, it is clearly necessary only to take those subsets that contain ϕ and X. Hence it is $\leq 2^{(2^n)-2}$. We must eliminate from here those subsets whose unions and intersections are not included in the topology. When $n = 1$ we get only 1 as is given by the upper bound. When $n = 2$ we get 4 and the formula gives $2^{4-2} = 2^2 = 4$ as the upper bound. However, for $n = 3$ the upper bound given is $2^{8-2} = 2^6 = 64$ of which only 29 are topologies (see the discussion at the end of Example 3 in 2.2). Of the remaining 35, 16 are excluded because the property of unions is not satisfied, 16 because the property of intersections is not satisfied and 3 because neither is satisfied. For $n = 4$ the total number of subsets is $2^{(2^4)} = 2^{16} = 65,536$ of which $2^{14} = 16,384$ have ϕ and X in them. There will be topologies which are non-trivial with 14 elements in them, for example. It is by no means easy to enumerate all possible topologies here, or to verify that they satisfy all the properties. As soon as we go to medium size sets with, say, 10 elements it becomes impossible. The total number of subsets not containing ϕ and X, is more than 10^{307}. The topology may have millions or billions, or more, sets in it. It would be impossible to distinguish between two specifications of the topology by enumeration in such cases.

The method of dealing with the problem is essentially that used when dealing with vectors in a finite dimensional space. There are infinitely many possible vectors. However, they can always be written in terms of a finite set of vectors. If the dimension of the space is n then a set of n or more vectors can be chosen such that every vector can be written as a linear combination of the given set of vectors. In fact all we require is that the components of the vectors, in some coordinate system, give a matrix of rank n. Such a system of vectors is called a basis. Generally a minimal set of basis vectors is used, i.e., the number of vectors is equal to the dimension of the space. Of course even the minimal basis is non-unique and there are infinitely many bases. Nevertheless one can convert from one basis to another and determine the components of the vector in various bases. As will appear shortly, the same sort of statements apply to the topological extension of the concept of a basis.

In a topological space (X, τ) a *basis* for τ, generally denoted by β, is a collection of subsets of X such that every element of τ, other than ϕ, can be expressed as a union of elements of β. For example, in τ_i the basis $\beta = \tau_i \smallsetminus \{\phi\} = \{X\}$. However, in τ_d, if $X = \{a_j\}(j = 1, \cdots, n)$ then

$\beta = \{\{a_j\}\}$ as every element of τ_d can be constructed from $\{a_1\}, \cdots, \{a_n\}$ and their unions. Thus the basis has n elements in it. Notice that this is an enormous reduction from the number of elements in this topology, namely 2^n. Further, if $|X| = \aleph_n$ then $|\beta| = \aleph_n$ for τ_d while $|\tau_d| = \aleph_{n+1}$. Clearly, there must always exist a basis in any topological space, since $\beta \subseteq \tau$ and we can at least take $\beta = \tau \smallsetminus \{\phi\}$. Let us, as usual, consider particular examples.

Example 1. Consider $(\mathbb{R}, \tau_u)$. Here $\beta_i = \{(a,b)|a,b \in \mathbb{Z}\}$ is *not* a basis as we have $(-\frac{1}{2}, \frac{1}{2}) \in \tau_u$ and there is no way to obtain it as a union of sets like $(-2,2), (-2,1), (-2,0), (-2,-1), (-1,2), (-1,1), (-1,0), (0,2), (0,1), (1,2)$, etc. However, $\beta = \{(a,b)|a,b \in \mathbb{R}\}$ is obviously a basis as $\beta = \tau_u \smallsetminus \{\phi\}$. Further, we can use $\beta_r = \{(a,b)|a,b \in \mathbb{Q}\}$ as a basis. This is true because every real number can be stated an interval that can be written as an infinite (or finite) series of rational numbers. Thus we can take an infinite union of sets in β_r to obtain any given set in τ_u. Again we have the basis $\beta_a = \{(a,b)|a,b \in \mathbb{A}\}$. Another choice of basis is $\beta_c = \{U|U \in \tau, |U'| \in \mathbb{N}\}$. The elements would, here, be specified by their complements which would be closed sets and finite, e.g., $U' = \{0\}$, $U' = \{e, \pi\}$ and so on. This fact is very easy to verify.

Example 2. If $X = \{a,b,c\}$ and $\tau = \{\phi, \{a\}, \{b\}, \{a,b\}, X\}$ then $\beta = \{\{a\}, \{b\}, X\}$ is a basis while $\tau \smallsetminus \{\phi\}$ is another basis. For $X = \{a,b,c,d\}, \tau = \{\phi, \{a\}, \{b\}, \{a,b\}, \{c,d\}\}, X\}$, $\beta = \{\{a\}, \{b\}, \{c,d\}\}$ is a basis for τ.

Example 3. In $(\mathbb{R}, \tau_c), \beta = \{\{x\}'|x \in \mathbb{R}\}$ is *not* a basis. The reason may be seen from the fact that
$$\mathbb{R} \smallsetminus \{x_1, x_2\} \neq (\mathbb{R} \smallsetminus \{x_1\}) \cup (\mathbb{R} \smallsetminus \{x_2\}) = \mathbb{R} \smallsetminus (\{x_1\} \cap \{x_2\}). \tag{2.42}$$
More generally, the generic open set cannot be written in terms of this set, as
$$\mathbb{R} \smallsetminus \{x_1, \cdots, x_n\} \neq (\mathbb{R} \smallsetminus \{x_1\}) \cup \cdots \cup (\mathbb{R} \smallsetminus \{x_n\}). \tag{2.43}$$
It is worthwhile to ponder how one might try to construct a non-trivial basis, i.e., $\beta \neq \tau \smallsetminus \{\phi\}$, for this topology. Would it be easier to use closed sets for the purpose, i.e., use the dual concept to form a basis?

2.11 Criteria for Topological Bases

We have seen that a given collection of open sets need not be a basis for a topology and that the basis is not unique. How, then, can one verify whether the collection is a basis or not? One way would be to verify by direct enumeration that the union of all elements of β lie in τ and that every element of τ is a union of elements of β. This method could only be used in the simplest cases, when the number of elements of β, and τ, is sufficiently small. In these cases, of course, we do *not need* to use a basis. Generally, however, this method would be impractical, at the very least, and may be totally impossible to use. As such we need alternative criteria to determine if a given collection is a basis. In what follows we proceed to look for precisely such criteria.

Theorem 2.11. In a topological space (X, τ), β is a basis iff $\forall x \in X$ and every $U \in \tau$ s.t. $x \in U, \exists B \in \beta$ s.t. $x \in B \subseteq U$.

Proof. For the first part suppose β is a basis. Consider $x \in U \in \tau$. By definition $U = \bigcup_{B \in \beta} B$ for some given Bs. Hence $x \in B$ for some $B \in \beta$ and it is clear that $B \subseteq U$. Conversely, assume that $\forall x \in U \in \tau \; \exists \; B \in \beta$ s.t. $x \in B \subseteq U$, then for each $x \in U$ we can find an open nbd, B_x, of x lying in β. Thus $U = \bigcup_{x \in U} B_x$. Thus every U can be expressed as a union of Bs and hence β is a basis. □

Corollary 2.11.1. If β is a basis in (X, τ) then:

(1)
$$\bigcup_{B \in \beta} B = X; \quad \text{or} \tag{2.44}$$

(2)
$$(\forall B_1, B_2 \in \beta \text{ and } x \in B_1 \cap B_2) \; \exists \; (B_3 \in \beta \text{ s.t. } x \in B_3 \subseteq B_1 \cap B_2). \tag{2.45}$$

Proof. These facts are immediately obvious from the above theorem. Notice that unions of elements of the basis do not have to be given in the basis but intersections do. □

Theorem 2.12. For a given space X, a collection β satisfying the properties of Corollary 2.11.1 uniquely specifies the topology.

Proof. Define the set
$$\tau = \{\phi, U | U = \bigcup_i B_i \text{ for some } B_i \in \beta\}. \tag{2.46}$$

By Eq. (2.44), $X \in \tau$. Also, it is obvious that if $B_1, B_2 \in \tau$ then $B_1 \cup B_2 \in \tau$. From Eq. (2.45) we see that $B_1 \cap B_2 \in \tau$. We could extend the former requirement for arbitrarily many elements of β as regards unions by the definition and finitely many intersections by applying Eq. (2.45) pairwise repeatedly. Hence the set defined by Eq. (2.46) satisfies all the properties of a topology. This topology is uniquely defined by the basis β. Note that β is *not* uniquely specified by a given τ but τ is uniquely specified by a given β. Thus if τ_1 and τ_2 have the same basis they must be the same. □

Example 1. If $X = \{a, b, c, d\}$ and $\beta = \{\{a\}, \{b\}, \{a, c\}, \{a, c, d\}\}$. Then $\phi, \{a\} \cup \{b\} = \{a, b\}, \{b\} \cup \{a, c\} = \{a, b, c\}$ and X also lie in τ. Thus

$$\tau = \{\phi, \{a\}, \{b\}, \{a, b\}, \{a, c\}, \{a, b, c\}, \{a, c, d\}, X\} \tag{2.47}$$

Example 2. For any X, $\beta_d = \{\{x\} | x \in X\}$ is the basis for the discrete topology and $\beta_i = \{X\}$ for the indiscrete topology, while for $X = \mathbb{R}$, $\beta_u = \{(a, b) | a, b \in \mathbb{R}\}$ is a basis for the usual topology on $\mathbb{R}$. It is easy to check that another basis for the usual topology is provided by $\beta = \{(x - \frac{1}{n}, x + \frac{1}{n}) | x \in \mathbb{R}, n \in \mathbb{N}\}$.

It is to be noted that for any topology $\tau, \tau_i \subseteq \tau \subseteq \tau_d$. It may not generally be possible to compare two topologies τ_1 and τ_2. However, if $\tau_1 \subsetneq \tau_2$ we say that τ_1 is *coarser* and τ_2 is *finer*, as the latter allows more topological structure to appear in the space which the former does not. Coarser and finer topologies are also called *smaller* and *larger* respectively. Clearly τ_i is the coarsest and τ_d the finest topology possible.

Theorem 2.13. Let β_1 and β_2 generate the topologies τ_1 and τ_2 respectively, on X. Then τ_2 is finer than τ_1 iff $\forall B_1 \in \beta_2$ s.t. $x \in B_1, \exists B_2 \in \beta_2$ s.t. $x \in B_2 \subseteq \beta_1$.

Proof. The essential point is that if τ_1 is coarser than τ_2 it has fewer elements than τ_2. Hence each of the elements of τ_2 should, in some sense, be "smaller than" the corresponding elements of τ_1. Only so could there be more "pieces" of the whole. Formally, the necessity is proved by noting that $B_1 \in \tau_1 \subsetneq \tau_2$ and so $B_2 \in \tau_2$. By Theorem 2.12, then, for any $x \in X$ we can find B_2 s.t. $x \in B_2 \subseteq B_1$. For sufficiency consider $x \in U \in \tau_1$. Since β_1 generates a topology, for any $x \in U, \exists B_1 \in \beta_1$ s.t. $x \in B_1 \subseteq U$ and by our assumption $\exists B_2 \in \beta_2$ s.t. $x \in B_2 \subseteq B_1$. Hence $x \in B_2 \subseteq U$. Thus $U \in \tau_2$. Hence $\tau_1 \subseteq \tau_2$. $\square$

2.12 Local Bases

In general one expects the dimension of a vector space to remain invariant from point to point. However, the space may become degenerate somewhere and become lower dimensional (for example). In that case the set of basis vectors may become smaller at some points than at others. To deal with such situations (and for other purely geometrical reasons) we set up the vector basis locally at each point. If the basis can be consistently transferred from point to point it is a coordinate basis and the components of the vector can be consistently dealt with, in the corresponding coordinates, using tensor methods. We want to extend this concept of a basis at a point to more general topological spaces. A *basis at a point*, or a *local basis*, at $x \in X, \beta_x$, is a collection of open *nbds* of x such that every *nbd* of x contains some element of β_x, i.e., $\forall x \in X, \beta_x = \{\eta(x) | \exists B \in \beta \text{ s.t. } B \subseteq \eta(x)\}$. Clearly $\mathfrak{N}(x)$ is a local basis. Also there exists at least one local basis, namely $\{X\}$. As examples, in $(X, \tau_d), \beta_x = \{\{x\}, \{x, a_i\} | a_i \in X \; \forall i \in I\}$ is a local basis. In $X = \{a, b, c\}$, $\tau = \{\phi, \{a\}, X\}, \beta_a = \{\{a\}, X\}, \beta_b = \{X\}$. Notice that β_x does not have to contain all open sets containing x. All that is required is that there be at least one open set containing x and the others can be generated from it.

Theorem 2.14. For a local basis, β_x, of a point $x \in X$,
(i) x belongs to every member of β_x
(ii) $B_1, B_2 \in \beta_x \implies \exists B_3 \in \beta_x$ s.t. $B_3 \subseteq B_1 \cap B_2$
(iii) $y \in B_1 \in \beta_x \implies \exists B_2 \in \beta_y$ s.t. $B_2 \subseteq B_1$.

The proofs of the above statements are very simple. They are left as an exercise.

The utility of the concept of a local base becomes apparent from the following observation. A collection of subsets of X, say G, s.t. $\forall x \in G \; \exists B_x \in \beta_x$ s.t. $B_x \subseteq G$ and including ϕ, is a topology on X, where β_x is a local base at x. Thus, from a local base at each point we can construct a topology on the entire space. Further $\beta = \{\beta_x\}$ over $x \in X$ is a basis for this topology.

If there exists a countable local base at each point of a space, that space is said to be *first countable*. If there is a countable basis for its entire topology the space is said to be *second countable*, or to satisfy the *second axiom of countability*. Clearly, every second countable space is first countable but the converse need not be true. For example $(\mathbb{R}, \tau_c)$ is first countable but not second countable. These concepts will be extensively used later when dealing with metric spaces.

Theorem 2.15. A second countable space is separable.

Proof. Given the basis $\beta = \{B_n | n \in \mathbb{N}\}$, which is countable, select a point a_n from each B_n. Consider any arbitrary $x \in X$ and $\eta(x) \in \mathfrak{N}(x)$ s.t. $\eta(x) \in \tau$. As β is a basis we can find $B_m \in \beta$ s.t. $B_m \subseteq \eta(x)$. Hence $a_m \in \eta(x)$. Thus every nbd $\eta(x)$, of x, contains a point of the set $A = \{a_n | n \in \mathbb{N}\}$. Hence A is dense in X. Thus (X, τ) is separable. □

Note that the converse of Theorem 2.15 is not generally true. For instance $(\mathbb{R}, \tau_c)$ is separable, since $\overline{\mathbb{N}} = \mathbb{R}$ in that topology, but it is not second countable. It will be shown later that a metric space is separable iff it is second countable.

Example 1. For $X = \mathbb{R}$ consider the A-inclusion topology with $A = \{x\}$ for some $x \in \mathbb{R}$. Here, the local base at x is $\{x\}$ and at any other point we can choose the local base to be $\{\mathbb{R}\}$. In this topology $\overline{\{x\}} = \mathbb{R}$. Hence $\{x\}$ is dense in $\mathbb{R}$ and so $\mathbb{R}$ is separable. Thus $\mathbb{R}$ is first countable but not second countable.

Example 2. Consider $\mathbb{R}$ with a topology generated by the basis $\beta = \{[a, b) | a \in \mathbb{R}, b \in \mathbb{Q}\}$. This is called the *Songfrey line*. $\mathbb{R}$ remains separable as $\mathbb{Q}$ remains dense in $\mathbb{R}$. It is first countable because the sets $[a, r)$ form a local basis at a as r ranges through all rationals greater than a. However, it is not second countable. It would be interesting to verify this fact.

2.13 Relative or Induced Topologies

Starting with a topological space (X, τ), we often need to limit our attention to a subspace of it. Suppose the subset we need to limit to is A, it is obvious that we cannot limit the topological space naively to (A, τ), as there will be elements of τ not contained in A. Thus we will need to modify the topology to those open sets that are contained in A, to obtain the topology τ_A which the subspace inherits from the original space. This *inherited*, or *induced*, or *relative topology* can be specified by

$$\tau_A = \{A \cap U | U \in \tau\}, \qquad (2.48)$$

for any $A \subseteq X$. It is obvious that $\phi, A \in \tau_A$ since $A \cap \phi = \phi$ and $A \cap X = A$. Further the union and finite intersections are contained in τ_A as

$$\bigcup_{i \in I}(A \cap U_i) = A \cap \left(\bigcup_{i \in I} U_i\right); \qquad (2.49)$$

$$\bigcap_{i=1}^{n}(A \cap U_i) = A \cap \left(\bigcap_{i=1}^{n} U_i\right). \qquad (2.50)$$

By a subspace of a given space we will always mean the subset with the relative topology as a topological space.

One would naturally expect that the closed sets of X would also be inherited by A. (Of course A could have been open, closed, both or neither open nor closed in (X, τ) but would be open and closed in (A, τ_A).) This expectation holds true as we see in the following theorem.

Theorem 2.16. If A is a subspace of X then $B \subseteq A$ is closed in subspace A iff $B = A \cap F$ for some closed set $F \subseteq X$.

Proof. Obviously, in A, $A \cap U' = (A \cap U)'$ as $(A \cap U)' = \phi \cup U'$ $\forall U \subseteq A$. Thus, since B is closed in A iff $A \smallsetminus B$ is open in A we must have $A \smallsetminus B = A \cap U$ for some $U \in \tau_A$. But, if $U \in \tau_A$ then $U \in \tau$ by definition. Thus $B = A \cap U' = A \cap (A \smallsetminus U) = A \cap (X \smallsetminus U)$ as $A \cap X = A = A \cap A$. Hence U' is closed in X. Notice that all open (and closed) sets in τ which are contained in A are open (and closed) in τ_A, but that there are additional open (and closed) sets in τ_A coming from subsets of sets contained in X that were not open or closed in τ. □

When considering a subspace we will need to be able to specify a basis for the relative topology. One may again expect that the part of the basis of τ contained in A would become the basis for τ_A. One would again be right. This is shown in the following theorem.

Theorem 2.17. If β is a basis for τ then a basis for τ_A is given by

$$\beta_A = \{A \cap B | B \in \beta\}. \tag{2.51}$$

Proof. Consider $X \in A \cap U$ for some $U \in \tau$. Since β is a basis $\exists B \in \beta$ s.t. $x \in B \subseteq U$. Thus $x \in A \cap B \subseteq A \cap U$. Hence β_A is a basis for τ_A by Theorem 2.11. □

One might further expect that if a set is open or closed in A it would also be open or closed in X. If one did, one would be wrong. A moment's reflection provides counterexamples. A, itself, is open in A but not necessarily in X. Also, consider $(\mathbb{R}, \tau_u)$ as the original topological space and take $[0, \infty)$ as A. Then $[0, a) \in \tau_A$ (for $a > 0$) but $[0, a) \notin \tau_u$. The reasonable expectation would be that open subsets in A are guaranteed to be open in X if A is open in X. The validity of this expectation is obvious as the intersection of two sets, open in A, will also be open in X. The dual statement also obviously holds, i.e., a subset of A is guaranteed to be closed in X if A is closed. Of course, we can find subsets of A that are open or closed in both A and X regardless of whether A is open or closed in X. In the above example (a, b) with $b > a > 0$ and $[a, b]$ are open and closed, respectively, in both A and X. Notice that in the above example $[0, a)$ is a nbd of 0 in A but not in X. Thus we need to specify nbds in the subspace. Generally, $\eta_A(x) = \eta(x) \cap A$. In the above example $(-\epsilon, a) \cap [0, \infty) = [0, a)$. The proof of the general statement is left as a simple exercise. Interiors and closures in A are inherited from X in the same way also. Thus $int_A(B) = int_X(B) \cap A$ and $\overline{(B)}^A = \overline{(B)}^X \cap A$, where "$\overline{}^A$" and "$\overline{}^X$" stands for "closure relative to τ_A or τ_X". Correspondingly, by definition, x is a limit point of a set B in the subspace iff it is a limit point of B in the original space.

Example 1. From $(\mathbb{R}, \tau_u)$ we can take $\mathbb{Z} \subseteq \mathbb{R}$ and look for the induced topology. Consider $U = (n - \frac{1}{2}, n + \frac{1}{2}) \in \tau_u, n \in \mathbb{N}$. Then $\mathbb{Z} \cap U = \{n\} \in \tau_\mathbb{Z}$. Hence $\tau_\mathbb{Z} = \{n | n \in \mathbb{Z}\}$. Hence the relative topology here is the discrete topology. Similarly, consider $\mathbb{Q} \subseteq \mathbb{R}$. Let $(r_a, r_b) \in \tau_u$. Then $(r_a, r_b) \cap \mathbb{Q} \in \tau_\mathbb{Q}$. Thus $\tau_\mathbb{Q} = \{(q_a, q_b) | q_a, q_b \in \mathbb{Q}, q_a \geq r_a$ (the equality holding if $r_a \in \mathbb{Q}$) and $q_b \leq r_b\}$.

Example 2. For a finite subspace of a co-finite space the inherited topology is discrete. The argument is the same as before and can be thought of more concretely by taking $X = \mathbb{R}$. For an infinite subspace the topology induced is still co-finite. To see this fact notice that

$$\tau_A = \{Y = A \cap U | U \in \tau_c\}, \tag{2.52}$$

i.e., Y is A if $A \subseteq U$, is ϕ whenever $A \cap U = \phi$, and U if $U \subseteq A$, apart from the case when A and U have a partial overlap. Thus (if $U \subseteq A$), $Y' = A \smallsetminus Y = A \smallsetminus U = U'$ is finite. If $A \subseteq U$ then, of course, $Y' = A \smallsetminus A = \phi$. Similarly, in the other cases Y' remains finite. Hence the topology is co-finite. It is a useful exercise to show that in the relative topology on $\mathbb{Q}$, inherited from the usual topology on $\mathbb{R}$,

$$(\overline{\mathbb{Z}})^{\mathbb{Q}} = \mathbb{Z}, \ (\overline{\mathbb{N}})^{\mathbb{Q}} = \mathbb{N}, \ \overline{([0,1])}^{\mathbb{Q}} \neq [0,1]. \tag{2.53}$$

2.14 Exercises

1. Let X be a topological space. Show that a subset U of X is open iff $A \cap U = \phi$ implies that $\bar{A} \cap U = \phi$ for each subset A of X.

2. Find a topological space X which has neither the discrete nor indiscrete topology in which subsets are open iff they are closed. (Hint: Let $X = \{a, b, c, d\}, \tau = \{\phi, X, \{a, b\}, \{c, d\}\}$.)

3. Let A and B be two subsets of a topological space X. Prove that:

   (a) $int(A \cap B) = int(A) \cap int(B)$;

   (b) $int(A) \cup int(B) \subseteq int(A \cup B)$;

   (c) $\overline{A \cup B} = \bar{A} \cup \bar{B}$;

   (d) Give an example of a topological space where $int(A) \cup int(B) \neq int(A \cup B)$ does not hold. (Hint: Consider $\mathbb{R}$ with respect to usual topology and take $A = \mathbb{Q}$, $B = \mathbb{Q}'$ then $int(\mathbb{Q}) \cup int(\mathbb{Q}') = \phi$ but $int(\mathbb{Q} \cup \mathbb{Q}') = \mathbb{R}$.)

4. Let X be any set endowed with the discrete topology. Show that every point is an isolated point of X.

5. Let X be given the indiscrete topology. Show that every point of X is a limit point for every subset of X.

6. Let X be a topological space. Show that if U is open in X and A is closed in X, then $U \smallsetminus A$ is open in X, and $A \smallsetminus U$ is closed in X.

7. Give an example of an infinite family $\{F_\alpha\}$ of closed sets in X whose union, $\bigcup_{\alpha \in I} F_\alpha$, is not closed.

8. Let $X = \{a, b, c\}$ and $\tau = \{\phi, X, \{b\}, \{c\}, \{b, c\}\}$ be a topology on X. Find the derived sets of all subsets of X.

9. Let X be an infinite set with the co-finite topology. Find the derived set, interior and closure for each of the subsets of X. (Hint: Let A be any subset of X, then either A is finite or infinite. When A is finite, then $A = \bar{A}$ and $int(A) = \phi$ and $\mathfrak{D}(A) = \phi$ as each point is an isolated point. Discuss the same when A is infinite.)

10. A space X is called a *door space* iff each subset of X is either open or closed. Find a door space that does not have either the discrete or the indiscrete topology. (Hint: Let $X = \{a, b, c\}$ and $\tau = \{\phi, X, \{a\}, \{b\}, \{a, b\}\}$ be a topology on X. Then X is a door space.)

11. Show that $int(A') = X \smallsetminus \bar{A}$ and $X \smallsetminus int(A) = \overline{(X \smallsetminus A)}$.

12. A subset A of a topological space is called a *perfect set* if $A = \mathfrak{D}(A)$. Give an example of a perfect set. Show that a set is perfect iff it is closed and has no isolated points.

13. Let A be a subset of a topological space. A point $x \in X$ is called a boundary point or frontier point of A if every open set U containing x intersects both A and A'. Denote by $F_r(A)$ the set of all boundary points of A. Show that

    (a) A is closed iff it contains its boundary, i.e., $A = A \cup F_r(A)$;
    (b) $F_r(A)$ is always a closed set;
    (c) $F_r(A) = \phi$ iff A is both open and closed;
    (d) A is open iff $A \cap F_r(A) = \phi$; A is closed iff $F_r(A) \subseteq A$.

14. A subset A of a topological space X is said to be nowhere-dense in X if $int(\bar{A}) = \phi$.

    (a) Show that the boundary of a closed set is nowhere-dense in X.
    (b) Prove that A is nowhere-dense in X iff its complement A' is everywhere dense.

15. Show that the co-finite topology on a finite set X coincides with the discrete topology on X.

16. Is there a set on which the discrete and the indiscrete topologies coincide? If so, determine it.

17. Find the boundary of each subset of X, where X is any set endowed with the co-finite topology. Also show that each point of X is a limit point of any infinite subset of A of X. Hence prove that if A is countably infinite then it is separable, i.e., $\bar{A} = X$.

18. Consider $\mathbb{R}$ with respect to the usual topology. Does there exist a countable subset A of $\mathbb{R}$, other than Q s.t. $\bar{A} = \mathbb{R}$? Give reasons.

19. Show that with respect to the discrete topology $\mathbb{R}$ is nonseparable.

20. Show that $F_r(\mathbb{Q}) = \mathbb{R}$ and $F_r(\mathbb{Q}') = \mathbb{R}$, $F_r(\mathbb{R}) = \phi$. This also means that $F_r(F_r(\mathbb{Q})) \neq F_r(\mathbb{Q})$.

Chapter 3

Metric Spaces

Though Topology is an extension of geometry to more unusual spaces, it needs to retain constant contact with the geometry of more "usual" spaces. Of course the question arises "What do we mean by usual spaces?" To start with, presumably, $\mathbb{R}^3 = \mathbb{R} \times \mathbb{R} \times \mathbb{R}$, with the usual topology, or its restriction to $\mathbb{R}^2$, or to $\mathbb{R}$. What, after all, could be more usual as a concept of "a space" than what is commonly called that? However, being mathematicians we develop this bad habit of generalizing everything and anything in sight. Why not, then, take $\mathbb{R}^n$ with the usual topology and geometry in it? When $n = 1, 2, 3$ we get the "very usual" space and otherwise a relatively "less usual" space, but still not very different. This is the Euclidean space. We have a slight problem of visualizing 4 or more dimensions as we can not physically construct 4 or more mutually orthogonal lines. However, by suppressing some dimensions at a time and looking at analogous lower dimensional spaces, we can obtain a fairly good comprehension of this "less usual" space.

Instead of the above generalization, with its concomitant problem of visualizing higher dimensions, why not restrict from $\mathbb{R}^3$ to some subspace defined by a relationship between the three variables, e.g.,

$$x^2 + y^2 + z^2 = a^2, \qquad (3.1)$$

(which is a sphere)? The geometry here is different from the geometry on $\mathbb{R}^2$ in that lengths along the surface of the sphere can not generally be expressed as the sum of the squares of two coordinates defined on the sphere. If one uses spherical polar coordinates on the surface of the sphere (θ, φ), the arc length for a displacement $d\theta$ and $d\varphi$ is

$$ds^2 = a^2(d\theta^2 + \sin^2\theta d\varphi^2), \qquad (3.2)$$

as opposed to the distance defined on a Euclidean plane in Cartesian coordinates

$$ds^2 = dx^2 + dy^2. \qquad (3.3)$$

You may ask why one should not use Cartesian coordinates on the surface of the sphere. The answer is that you can only use three Cartesian coordinates *constrained* by Eq. (3.1) for the surface of the sphere. Thus we will not have three independent variables and will consequently be unable to use the procedure of partial differentiation. To use calculus on the space we need independent variables as coordinates. On a curved surface we can not use Cartesian coordinates as they will either not be independent or not give the length correctly.

On an arbitrary surface, parameterized by two variables (say u and v) we can construct tangent vectors along the parameter lines $v =$ constant and $u =$ constant at the point where these lines intersect. So long as the two vectors do not coincide we can use these as basis vectors for the surface. If the position vector is denoted by $\mathbf{x}(u,v)$ and the partial derivatives by $\mathbf{x}_u$ and $\mathbf{x}_v$, we can compute $\mathbf{x}_u \cdot \mathbf{x}_u$, $\mathbf{x}_u \cdot \mathbf{x}_v$ and $\mathbf{x}_v \cdot \mathbf{x}_v$. For a displacement by (du, dv) the arc length is given by

$$\begin{aligned} ds^2 &= d\mathbf{x} \cdot d\mathbf{x} = (\mathbf{x}_u du + \mathbf{x}_v dv) \cdot (\mathbf{x}_u du + \mathbf{x}_v dv) \\ &= \mathbf{x}_u \cdot \mathbf{x}_u du^2 + 2\mathbf{x}_u \cdot \mathbf{x}_v dudv + \mathbf{x}_v \cdot \mathbf{x}_v dv^2. \end{aligned} \qquad (3.4)$$

This may be expressed more conveniently by putting $u = u_1$ and $v = u_2$, $\mathbf{x}_u \cdot \mathbf{x}_u = g_{11}$, $\mathbf{x}_u \cdot \mathbf{x}_v = g_{12}$, $\mathbf{x}_v \cdot \mathbf{x}_v = g_{22}$. Then

$$ds^2 = \sum_{i,j=1}^{2} g_{ij}(u_k) du_i du_j. \qquad (3.5)$$

The g_{ij} here are called *metric coefficients* as they define the length measure. The arc length squared is called the *metric* and the matrix of metric coefficients is called the *metric tensor* (when conceived of as written independent of the choice of parameter). This generalization was developed by various people, but the lion's share of credit must clearly go to Gauss.

Riemann extended this generalization to higher dimensional spaces, thus enormously increasing the power and general applicability of geometry. He investigated spaces with the metric defined by

$$ds^2 = \sum_{i,j=1}^{n} g_{ij}(u_k) du_i du_j, \qquad (3.6)$$

such that the determinant of the matrix, g_{ij}, remains positive definite, so that the arc length square can be guaranteed to be positive definite, i.e., never negative and zero only when $d\mathbf{u} = \mathbf{0}$. Though Eq. (3.6) hardly seems different from Eq. (3.5), merely replacing 2 by n, it is a major leap taken. The increase of the number of variables increases the complexity of the geometry enormously. We will not pursue these developments here, however. The interested reader may consult [Qadir, 2020].

Our purpose is to follow the more abstract topological approach to develop geometry. We will not start with the assumption that partial differentiation can be performed. We will not even assume that calculus can be used. In fact the basis of calculus must lie in Topology as *it* can tell us which spaces satisfy the criteria for the use of calculus. How can we develop geometry without having a space on which calculus can be used? The answer is not to follow the path taken from Eq. (3.2) through Eq. (3.5). Instead, we take the essential definition of the arc length as *a priori* given by our free choice, not defined by partial derivatives. Again, there is no way we can talk of infinitesimal quantities in such an arbitrary space. Thus the metric must be defined more generally here.

A space with a metric defined on it, in this general way, is the subject of this chapter. We will see how the metric leads naturally to a topology. However, it is not necessary that a space with a given topology have a metric definable on it. Recall the earlier example of all people, regarded as economic entities (consumers, etc.) in Section 1.4. Call this set X and consider (X, τ_i). Clearly there can be no metric definable on this set of isolated points — where "no man is an island complete unto himself" [Donne and Lush, 1988].

Metric spaces will be needed for developing the ideas on continuity, limits, differentiability, etc., in a sufficiently arbitrary setting that we will be able to use calculus in spaces with cardinality $\aleph_2$. Since this is the cardinality of the space of all functions, we *need* to be able to develop the use of calculus for such spaces. That such a development is non-trivial becomes apparent when you try to define how a sequence of functions converges to one limiting function. This would be necessary if we want to differentiate a mapping from the space of functions into itself. Consider the functions $f(x), g(x), h(x)$ and $k(x)$ shown in Figure 3.1. How can we decide whether $g(x), h(x)$ or $k(x)$ is closer to $f(x)$? We need to define a measure for the difference and the measure may be non-unique. It would depend on the context as to what measure is defined.

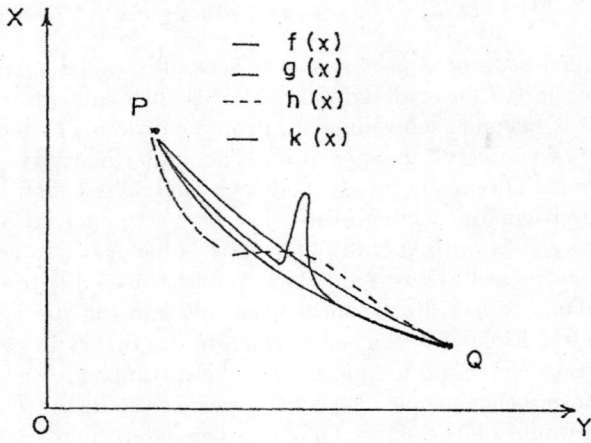

Figure 3.1: Given the curve from P to Q, $f(x)$, it is not possible to decide which of the curves from P to Q shown here ($g(x), h(x)$ or $k(x)$) comes closest to $f(x)$. Each, in its own way is closest.

Even in spaces which are continua, i.e., have cardinality $\aleph_1$, the measure of distance between points of the space may not be obvious. For example, consider the linear vector space of square $n \times n$ matrices. There is no clear definition of the difference between matrices tending to zero. For example if $A = \begin{pmatrix} 1 & 2 \\ 3 & 4 \end{pmatrix}, B = \begin{pmatrix} 1 & 2 \\ 3 & 7 \end{pmatrix}, C = \begin{pmatrix} 0 & 3 \\ 4 & 3 \end{pmatrix}$, is C closer to A or B closer to A? We need general methods of defining the "distance" between two points (i.e., elements of a space).

3.1 The Definition of a Metric

Though we want the measure of the distance to be arbitrary, we do not want it to be *totally* arbitrary. First and foremost it must be a number which can tend to zero as in a continuum. Hence it must be a real number. Also, the distance from a point to itself must be zero and to another must be the same as that from the other point to the first. Finally, we will require that the theorem of plane geometry, that the sum of two sides of a triangle is greater than the third side, should continue to hold for our arbitrary measure. Let us state these requirements more precisely.

A *metric*, ρ, on a space, X, is a mapping

$$\rho : X \times X \to \mathbb{R}, \qquad (3.7)$$

given by

$$\rho(x,y) = r \quad (x,y \in X,\ r \in \mathbb{R}), \qquad (3.8)$$

satisfying the following properties:

$$\begin{aligned}
(1)\quad & \rho(x,x) = 0,\quad \rho(x,y) > 0 \quad (y \neq x); & (3.9)\\
(2)\quad & \rho(x,y) = \rho(y,x); & (3.10)\\
(3)\quad & \rho(x,z) \leq \rho(x,y) + \rho(y,z). & (3.11)
\end{aligned}$$

This first property is called *positive definiteness*, the second is called *symmetry* and the third is called the *triangular inequality*. The ordered pair (X,ρ) is then called a *metric space*.

In some applications it becomes convenient to drop conditions (1) and (3). This is the case, for example, in the use of geometry in Special and General Relativity. In that case ρ would be called a *pseudo-metric*. The geometry defined by a metric is called Riemannian geometry, since it was developed by Riemann. Though the other geometry, defined by a *pseudo-metric* was developed by Lobachevsky, it is *not* generally known as Lobachevskian geometry, but rather as pseudo-Riemannian geometry, see Figure 1.1. (Presumably, this is due to some prejudice against the Russian mathematician.) In fact, in the special case relevant for Special Relativity, it is called a Minkowski geometry after Einstein's teacher who wrote the theory in geometrical terms.

There is, as pointed out earlier, no unique measure of distance in a space. Consider a person who keeps his distance from other people. For such a person the distance from himself would be zero and from all others would be the same. On the other hand, there may be someone who has strong family ties. Her distance from herself may be zero, from her parents, siblings and offspring next, from her husband a bit more, from her cousins still more, from her in-laws still more, and so on. Let us consider more usual (and more mathematical) examples. Our choice is particularly directed to those that are more commonly used.

Example 1. $\rho_d(x,x) = 0, \rho_d(x,y) = 1\ \forall x,y \in X$ is a metric on X. As will be seen later, it leads to the discrete topology and is, therefore, called the *discrete metric*.

Example 2. If $|X| > 1$ then $\rho(x,y) = 0\ \forall x,y \in X$ is *not* a metric on X.

Example 3. For $X = \mathbb{R}$, $\rho_u(x,y) = |x-y|\ \forall x,y \in \mathbb{R}$ is a metric as $\rho_u(x,x) = 0$ and $\rho_u(x,y) > 0$ for $y \neq x$. Also, $\rho_u(x,y) = \rho_u(y,x)$ by definition and $|x+y| = |x| + |y|$ if x,y have the same sign but $|x+y| < |x| + |y|$ if they have opposite signs. This is the *usual metric* on $\mathbb{R}$. Now define

$$\rho_n(x,y) = |x-y|/n,\quad (n \in \mathbb{N}). \qquad (3.12)$$

This is a "shrunken" interval measure. Similarly

$$\rho_m(x,y) = m\,|x-y|,\quad (m \in \mathbb{N}), \qquad (3.13)$$

is an "expanded" metric on $\mathbb{R}$.

Example 4. On $X = \mathbb{C} \ni z = z_1 + iz_2$ with $z_1, z_2 \in \mathbb{R}$, we have the *usual* or *standard metric*

$$\begin{aligned} \rho_u(x,y) &= |x-y| = |x_1 - y_1 + i(x_2 - y_2)| \\ &= \left[(x_1 - y_1)^2 + (x_2 - y_2)^2\right]^{\frac{1}{2}}. \end{aligned} \quad (3.14)$$

Instead we can define

$$\rho_c(x,y) = \frac{2|x-y|}{\sqrt{1+|x|^2} + \sqrt{1+|y|^2}}, \quad (3.15)$$

called the *chordal distance*.

Example 5. In a problem like making a 5 year plan for the economy, where one knows that the actual goal will probably not be achieved, one is primarily concerned with the extent to which one may fall short in any one of several variables. For example, one may want a certain rate of economic growth, a certain GNP (Gross National Product), a certain per capita income and a limit to income inequalities. It would not be so relevant to look for the difference between the desired and the actual value of all put together, but rather the one for which the difference is a maximum (and perhaps unacceptably high). Thus $X = \mathbb{R}^n$ and the metric, $\rho_1(x_i, y_i)$ $(i = 1, \cdots, n)$ is given by the *square metric*

$$\rho_1(\mathbf{x}, \mathbf{y}) = \max_{1 \leq i \leq n} \{|x_i - y_i|\}, \quad (3.16)$$

i.e., the magnitude of the largest component of the vector $\mathbf{y} - \mathbf{x}$, see Figure 3.2.

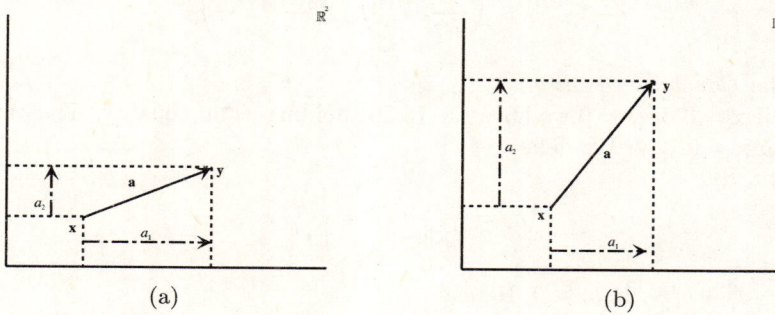

(a)   (b)

Figure 3.2: Consider $\mathbb{R}^2$ for the illustration of this metric. Two points are given by position vectors $\mathbf{x}$ and $\mathbf{y}$ and $\mathbf{a} = \mathbf{y} - \mathbf{x}$. In (a) $a_1 \gg a_2$ and hence $\rho(\mathbf{x}, \mathbf{y}) = a_1$ while in (b) $a_2 > a_1$ and hence $\rho(\mathbf{x}, \mathbf{y}) = a_2$. If the two were equal $\rho(\mathbf{x}, \mathbf{y}) = a_1 = a_2$. Notice that the "magnitude" so defined is not the usual expectation of $|\mathbf{a}|$.

Positive definiteness and symmetry is obvious. For the triangular inequality note that

$$\begin{aligned} |z_i - x_i| &\leq |y_i - x_i| + |z_i - y_i| \quad \forall i = 1, \cdots, n & (3.17) \\ &\leq \max_{1 \leq j \leq n} \{|y_j - x_j|\} + \max_{1 \leq k \leq n} \{|z_k - y_k|\} \forall i. & (3.18) \end{aligned}$$

Thus, taking the maximum over i and using Eq. (3.16)

$$\rho_1(\mathbf{x}, \mathbf{z}) \leq \rho_1(\mathbf{x}, \mathbf{y}) + \rho_1(\mathbf{y}, \mathbf{z}). \quad (3.19)$$

Instead one could add up all the component magnitudes.

$$\rho_2(\mathbf{x}, \mathbf{y}) = \sum_{i=1}^{n} |y_i - x_i|. \tag{3.20}$$

Again positive definiteness and symmetry are obvious and for the triangular inequality we can sum (3.18) over all i to obtain

$$\rho_2(\mathbf{x}, \mathbf{z}) \le \rho_2(\mathbf{x}, \mathbf{y}) + \rho_2(\mathbf{y}, \mathbf{z}). \tag{3.21}$$

The *usual metric* on $\mathbb{R}^n$, called the *norm*, is the geometrical length of the vector $\mathbf{y} - \mathbf{x}$ in an Euclidean space:

$$\|\mathbf{y} - \mathbf{x}\| \equiv \rho_u(\mathbf{x}, \mathbf{y}) = \left[\sum_{i=1}^{n} (y_i - x_i)^2\right]^{\frac{1}{2}}. \tag{3.22}$$

That this metric satisfies positive definiteness and symmetry requirements is obvious. Though the triangular inequality seems obvious it is not really so simple as merely to quote plane geometry theorems. We prove it with the help of two Lemmas.

Lemma 1.1: For $\mathbf{x}, \mathbf{y} \in \mathbb{R}^n, |\mathbf{x} \cdot \mathbf{y}| \le \|\mathbf{x}\| \cdot \|\mathbf{y}\|$ or

$$\sum_{i=1}^{n} |x_i y_i| \le \left(\sum_{i=1}^{n} |x_i|^2\right)^{\frac{1}{2}} \left(\sum_{j=1}^{n} |y_j|^2\right)^{\frac{1}{2}}. \tag{3.23}$$

(This is called the *Cauchy inequality*.)
Proof: Clearly if $\mathbf{x} = 0$ or $\mathbf{y} = 0$ we have Eq. (3.23) holding as an equality. Thus we only need to consider the case $\mathbf{x} \ne 0 \ne \mathbf{y}$. We define

$$a_i = \frac{|x_i|}{\|\mathbf{x}\|}, \quad b_i = \frac{|y_i|}{\|\mathbf{y}\|}. \tag{3.24}$$

Now $(a_i - b_i)^2 = a_i^2 + b_i^2 - 2a_i b_i > 0$. Hence

$$a_i b_i < (a_i^2 + b_i^2)/2. \tag{3.25}$$

Using Eq. (3.24) in Eq. (3.25) and summing up over all i, we get

$$\sum_{i=1}^{n} |x_i y_i| / (\|\mathbf{x}\| \cdot \|\mathbf{y}\|) < \frac{1}{2}\left[\sum_{i=1}^{n} |x_i|^2 / \|\mathbf{x}\|^2 + \sum_{i=1}^{n} |y_i|^2 / \|\mathbf{y}\|^2\right]$$
$$= \frac{1}{2}(1+1) = 1. \tag{3.26}$$

Hence we see that inequality (3.23) holds.

Lemma 1.2: For $\mathbf{x}, \mathbf{y} \in \mathbb{R}^n$, $\|\mathbf{x} + \mathbf{y}\| \leq \|\mathbf{x}\| + \|\mathbf{y}\|$ or

$$\left(\sum_{i=1}^{n} |x_i + y_i|^2\right)^{\frac{1}{2}} \leq \left(\sum_{i=1}^{n} |x_i|^2\right)^{\frac{1}{2}} + \left(\sum_{i=1}^{n} |y_i|^2\right)^{\frac{1}{2}}. \tag{3.27}$$

(This result is known as *Minkowski inequality*.)

Proof: We have

$$\|\mathbf{x} + \mathbf{y}\|^2 = \sum_{i=1}^{n} |x_i + y_i|^2 = \sum_{i=1}^{n} |x_i + y_i| \cdot |x_i + y_i|$$

$$\leq \sum_{i=1}^{n} |x_i + y_i|(|x_i| + |y_i|) \tag{3.28}$$

by the triangular inequality on $\mathbb{R}$. Thus

$$\|\mathbf{x} + \mathbf{y}\|^2 \leq \sum_{i=1}^{n} |x_i + y_i||x_i| + \sum_{i=1}^{n} |x_i + y_i||y_i|$$

$$\leq \left(\sum_{i=1}^{n} |x_i + y_i|^2\right)^{\frac{1}{2}} \left(\sum_{j=1}^{n} |x_j|^2\right)^{\frac{1}{2}} + \left(\sum_{i=1}^{n} |x_i + y_i|^2\right)^{\frac{1}{2}} \left(\sum_{j=1}^{n} |y_j|^2\right)^{\frac{1}{2}}$$

by the Cauchy inequality. Hence

$$\begin{aligned}\|\mathbf{x} + \mathbf{y}\|^2 &\leq \|\mathbf{x} + \mathbf{y}\| \cdot \|\mathbf{x}\| + \|\mathbf{x} + \mathbf{y}\| \cdot \|\mathbf{y}\| \\ &= \|\mathbf{x} + \mathbf{y}\|(\|\mathbf{x}\| + \|\mathbf{y}\|).\end{aligned} \tag{3.29}$$

Hence the inequality (3.27) holds. This states that the metric defined by the norm satisfies the triangular inequality as we had asserted.

Example 6. Consider the space of all real valued and continuous functions defined on the closed interval $[a, b]$, denoted by $C([a, b])$. This was one of the problems we had mentioned earlier where there is no obvious metric but one is needed. The equivalent of the square metric, here is

$$\rho_1(f, g) = \max_{a \leq t \leq b} |f(t) - g(t)| \quad \forall f, g \in C([a, b]). \tag{3.30}$$

Here the discretely varying i, over the finite index set $1, \cdots, n$ is replaced by the continuously varying t, but there is no other difference. The extension is left as an exercise for you. The generalization of ρ_2 and ρ_u for functions involves integration. Notice that an $n \times n$ matrix could be thought of as an n^2 column vector for the purposes of defining matrices like ρ_1 and ρ_2 for them. Of course ρ_u could also be defined in the same way but one would have to bear in mind that the geometrical significance it carries for vectors would not hold for the metrics.

Example 7. Consider the sequence of real numbers

$$\mathbf{x} = (x_1, x_2, \cdots, x_i, \cdots), \qquad x_i \in \mathbb{R}, \ i \in \mathbb{N}. \tag{3.31}$$

It is said to be *bounded* if $\exists K$ s.t. $|x_i| \leq K \ \forall i \in \mathbb{N}$. The space of all bounded sequences has *its* equivalent of the square metric defined on it,

$$\rho_1(\mathbf{x}, \mathbf{y}) = \sup_i |y_i - x_i|, \qquad \forall \mathbf{x}, \mathbf{y} \in X, \tag{3.32}$$

where "sup" stands for *supremum*, by which we mean the least upper bound for the set. Correspondingly, the lower bound is called the *infimum* and written as "inf". The generalizations of ρ_2 and ρ_u may not always exist. For ρ_u to be defined we require that all series $\sum_{i=1}^{n} x_i^2$, of sequences $\mathbf{x}$ in the spaces, must converge. Such a space is denoted by ℓ_2. For ρ_2 to be definable it is necessary that the series $\sum_{i=1}^{n} |x_i|$ be absolutely convergent. These metrics should be verified to satisfy the desired properties.

3.2 The Topology Induced by a Metric

It seems intuitively obvious that once we have a metric on a space we should be able to provide an equivalent of the open interval on $\mathbb{R}$ and hence define a topology on the metric space. This is indeed the case. To do so we first need to provide the formal generalization of the open interval. On the metric space (X, ρ) we define the *open sphere centred at x*

$$S_r(x) = \{y \in X | \rho(x,y) < r\}. \tag{3.33}$$

Though X need not be $\mathbb{R}^3$, for $(\mathbb{R}^3, \rho_u)$ the name is obviously appropriate (see Figure 3.3). The sphere is never empty, as it always contains its centre, x. We could, correspondingly, have defined the *closed sphere*

$$\overline{S}_r(x) = \{y \in X | \rho(x,y) \leq r\}, \tag{3.34}$$

to provide the dual development of the topology.

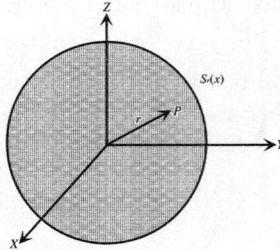

Figure 3.3: The open sphere in $\mathbb{R}^3$, $S_r(x)$, is simply the interior of the sphere of radius r, in the case of the usual metric. Here r is the magnitude of a typical radius vector, $\mathbf{r}$, from the centre O (corresponding to x) to a point P (corresponding to z such that $\rho(x,z) = r$).

Clearly, we can define open sets to be open spheres and their unions and finite intersections. Let us state this more formally. A subset U of X will be said to be *open* if $\forall x \in U \ \exists r \in \mathbb{R}$ s.t. $S_r(x) \subseteq U$. In other words, around each point of U we can draw an open sphere which is fully contained in U. The topology obtained by this definition is said to have been induced by the metric. Before going on to show that we do, indeed, have a consistent definition of a topology, let us consider some examples.

Example 1. For a space with the discrete metric defined on it $S_r(x) = \{x\}$ for $0 < r \le 1$ and $S_r(x) = X$ for $r > 1$.

Example 2. For $\mathbb{R}$ with the usual metric $S_r(x)$ is the usual open interval centered at x, $(x - r, x+r)$. If we *expand* the metric, so that $\rho_m(x,y) = m\,|x-y|$ then the open interval will *contract* as
$$S_r(x) = \{y \in \mathbb{R}|\ m\,|x-y| < r\} = (x - r/m,\ x + r/m) \qquad (3.35)$$
and *vice versa*, i.e., if $m < 1$ the sphere expands.

Example 3. Consider $\mathbb{R}^3$ with the square metric, ρ_1. The "open sphere" in this case is an *open square*
$$S_r(\mathbf{x}) = \left\{\mathbf{y} \in \mathbb{R}^2|\ \max\left(|x_1 - y_1|, |y_2 - x_2|\right) < r\right\}, \qquad (3.36)$$
which has side $2r$ (see Figure 3.4). Instead, using ρ_2 we obtain the open diamond (see Figure 3.5) given by
$$S_r(\mathbf{x}) = \left\{\mathbf{y} \in \mathbb{R}^2|\ |y_1 - x_1| + |y_2 - x_2| < r\right\}, \qquad (3.37)$$
while for ρ_u we get the open disc of radius r.

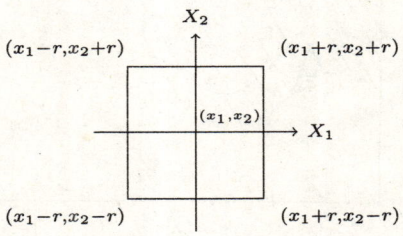

Figure 3.4: The "open sphere" for the square metric is the "open square" shown here. This is why the square metric is called "square".

Example 4. In the space $C([0,1])$, with the generalization of the square metric for functions, the open sphere is given as,
$$S_r(f) = \{g | g \in C([0,1]),\ \rho_1(f,g) < r\}, \qquad (3.38)$$
which is a band of width $2r$ centered about $f(t)$ as shown in Figure 3.6. With $\rho_2(f,g)$ defined by
$$\rho_2(f,g) = \int_0^1 |f(t) - g(t)|\,dt, \qquad (3.39)$$
and $\rho_u(f,g)$ defined by
$$\rho_u^2(f,g) = \int_0^1 [f(t) - g(t)]^2\,dt, \qquad (3.40)$$
we would get the corresponding "open spheres", neither of which is so easily represented.

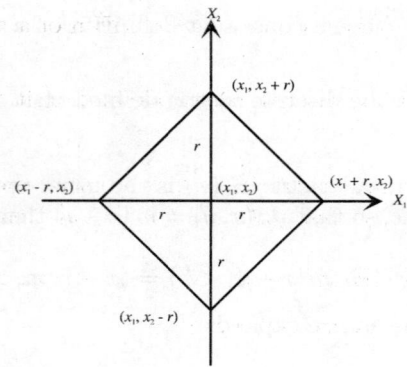

Figure 3.5: The metric ρ_2 gives the "open diamond" as it is bounded at the upper right by the line $y_1 + y_2 = x_1 + x_2 + r$, at the upper left by $y_1 - y_2 = r + x_2 - x_1$, at the lower right by $y_2 - y_1 = r + x_1 - x_2$ and at the lower left by $y_1 + y_2 = x_1 + x_2 - r$.

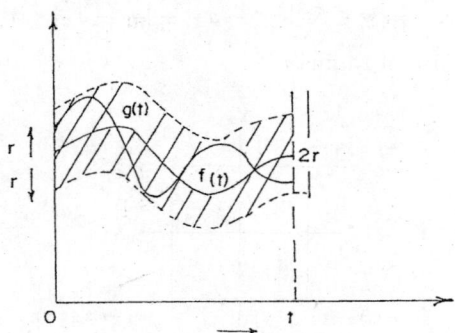

Figure 3.6: For a typical function $f(t) \in C([0, 1])$, $S_r(f)$ is the ribbon of width $2r$ through the centre of which runs $f(t)$, so that any $g(t) \in S_r(f)$ is such that $\rho(f, g) < r$. Notice that $g(t)$ must never touch the edge of the ribbon but can come arbitrarily close to it.

So far we have constructed $S_r(x)$ so that it plays the role of an "open set" as a generalization of the open interval for arbitrary metric spaces. We have not, as yet, shown that they satisfy all the requirements we had imposed on open sets. In particular, we have not demonstrated that the $S_r(x)$ are actually open sets or shown that the union of open spheres centered at different points gives any open set or that any open set can be expressed as a union of open spheres centered at different points. We now want to demonstrate that we can use the concept of open spheres to generate a topology. This result will be proved by a series of three theorems. In the first we will show that the open spheres satisfy the characterization of open sets that if η is an open *nbd* of x which contains y then $\exists \, U \subseteq \eta$ s.t. $y \in U \in \tau$. Here we will need to consider a sphere centered at x and show that any y in that sphere can be enclosed by an open sphere *centered at* y which lies in the sphere centered at x. It is obvious that if we had taken a closed sphere then this condition would not have been satisfied on the boundary, at least for the usual topology.

Metric Spaces

Theorem 3.1. In a metric space every open sphere is an open set.

Proof. For any $x \in X$ consider some $S_r(x)$. Now take an arbitrary point $y \in S_r(x)$. Since $y \in S_r(x), \rho(x,y) < r$. Choose $s = r - \rho(x,y)$. Clearly, $s > 0$. Now consider any $z \in S_s(y)$. Then

$$\rho(x,z) \leq \rho(x,y) + \rho(y,z) < \rho(x,y) + s = r. \tag{3.41}$$

Hence $z \in S_r(x) \ \forall z \in S_s(y)$. Hence $S_s(y) \subseteq S_r(x)$. Thus the open sphere satisfies the characterization of an open set in an arbitrary topological space and so an open sphere is an open set. $\square$

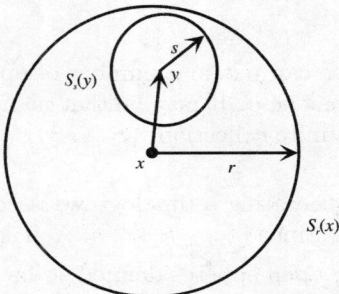

Figure 3.7: For a given x and r, $S_r(x)$ is given by the metric. For any $y \in S_r(x)$ we can find an s such that $S_s(y) \subseteq S_r(x)$.

This is illustrated in Figure 3.7. We must now proceed to show that any open set can be expressed as a union of open spheres and that any union of open spheres is an open set. This result is proved in the following theorem.

Theorem 3.2. In a metric space (X, ρ), a subset U is open iff it is a union of open spheres.

Proof. We start by proving the necessity of the condition. If U is an open set, since $S_r(x)$ is an open set (by the previous theorem), $\forall x \in U, \exists\, r > 0$ s.t. $x \in S_r(x) \subseteq U$. Taking the union of all open spheres centered at points in U we have $\bigcup_{\substack{x \in U \\ r > 0}} S_r(x)$ with r allowed to vary over those values for which $S_r(x) \subseteq U$ only. However, since $\bigcup_{x \in U} \{x\} = U$, clearly $\bigcup_{x \in U} S_r(x) \supseteq U$. Hence $\bigcup_{\substack{x \in U \\ r > 0}} S_r = U$. Thus every open set can be expressed as a union of open spheres. We now prove sufficiency. Consider a subset U of X, which can be expressed as a union of open spheres. Thus every $x \in U$ lies in some open sphere $S_r(y) \subseteq U$. Since $S_r(y)$ is an open set (by Theorem 3.1), $\exists\, S_r(x) \subseteq S_r(y)$ s.t. $S_r(x) \subseteq U$. Hence U is open by the characterization (referred to above) of open sets. $\square$

This point is illustrated in Figure 3.8.

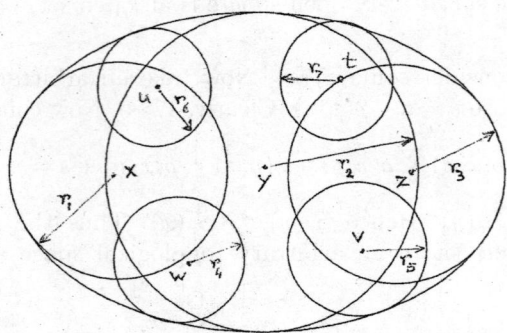

Figure 3.8: For any open set U we can put in a number of open spheres which overlap so that, between them, they cover every point of U. It may be that no finite collection of open spheres will suffice. In that case we take an infinite collection, $\{S_{r_i}(x_i)\}_{i \in I}$, s.t. $\bigcup_{i \in I} S_{r_i}(x_i) = U$.

To prove that we are indeed generating a topology we need to prove that ϕ and X are open sets in the sense of open spheres. Since $\bigcup_{\substack{x \in X \\ r > 0}} S_r(x) = X$ it is clear that X is open. That ϕ is open is slightly more subtle. The open spheres defined so far were non-empty. We said U was open if $\exists\, r > 0$ s.t. $S_r(x) \subseteq U$ $\forall x \in U$. In the case of the empty set there is no $x \in \phi$. Hence any r satisfies the condition. In other words the condition becomes irrelevant for ϕ. Thus ϕ is an open set in this definition (to the extent that ϕ is a set). In so far as ϕ was introduced for the sake of consistency, it is consistent to regard it as an open set here.

The final step remaining to be taken is to prove that the collection of open spheres provides a basis for a topology. We proceed, now, to prove this.

Theorem 3.3. In the metric space (X, ρ) the collection of open spheres

$$\beta = \{S_r(x) | x \in X, r > 0\}, \tag{3.42}$$

is a basis for a topology on X.

Proof. By Theorem 3.2, $\exists\, r > 0$ s.t. $x \in S_r(x) \subseteq U$, where $S_r(x) \in \beta$. This is the requirement for a basis. $\square$

We have, therefore, proved that a metric must always generate a topology. There can be topological spaces for which no metric can be defined consistently with the topology. However, there cannot be a metric space for which a topology cannot be defined. Thus topological spaces are more general than metric spaces. A topological space on which a metric can be defined is called a *metrizable space*. It is implicit here that the topology generated by the metric is essentially the topology of the space already given.

Example 5. The discrete metric on any set generates the discrete topology as $S_r(x) = X$ for $r > 1$. Thus $\beta = \{S_r(x) | x \in X\} = \{\{x\} | x \in X\}$, which is the basis for the discrete topology.

Metric Spaces

Example 6. The usual metric on $\mathbb{R}$ generates the usual topology as the basis is $\beta_u = \{(x-r, x+r) | x \in \mathbb{R}, r > 0\}$. On $\mathbb{R}^n$ it is $\beta_u = \{S_r(\mathbf{x}) | \mathbf{x} \in \mathbb{R}^n, r > 0\}$ where $S_r(\mathbf{x})$ is the n-dimensional open sphere.

It is not necessary that different metrics generate different topologies. Different bases generated by different metrics may produce the same topologies as we know that there is no unique base for a topology in general. The "open spheres" may be *geometrically distinct* objects that are *topologically equivalent*. Topological equivalence will be at the root of the next chapter and is discussed more fully in the next section.

3.3 Equivalent Topologies Generated by Metrics

Different metrics on the same space may or may not generate different topologies. For example, the usual metric generates the usual topology on $\mathbb{R}$ while the discrete metric generates the discrete topology. An example of different metrics generating the same topology is provided by the following theorem.

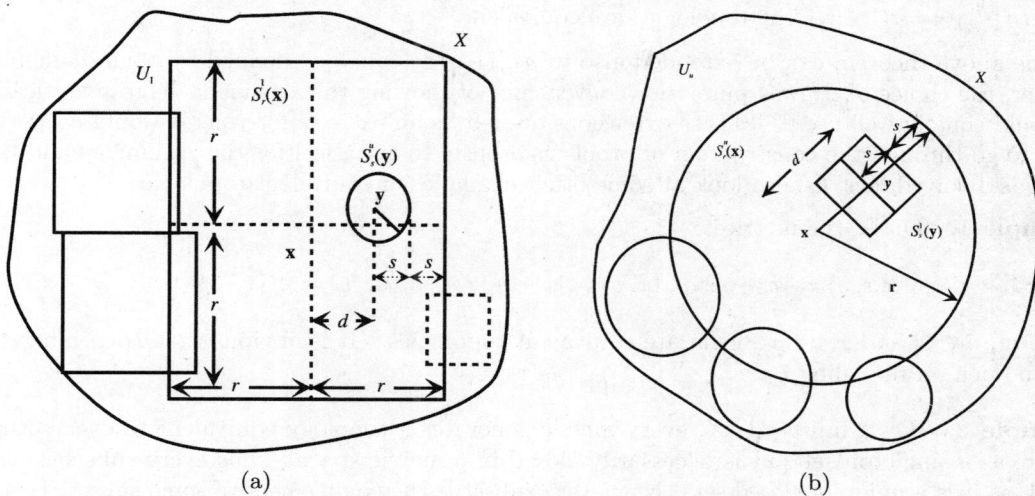

Figure 3.9: (a) A region $U_1 \subseteq X$ consisting of unions of open squares has a typical open square $S_r^1(\mathbf{x})$ centred at $\mathbf{x}$ with a ρ_1-size r. A point $\mathbf{y} \in S_r^1(\mathbf{x})$ is at ρ_1-distance d from $\mathbf{x}$. Take $s = (r-d)/2$. Then a disc $S_s^u(\mathbf{y})$ centred at $\mathbf{y}$ has minimum ρ_1-distance s from the boundary of the open square. Hence every point inside the open square can be enclosed in an open disc and so U_1 can be taken to be a union of open discs. (b) Conversely, a region $U_u \subseteq X$ is a union of open discs with a typical open disc $S_r^u(\mathbf{x})$. A point $\mathbf{y}$ is at distance d from $\mathbf{x}$ in the usual sense. Take $s = (r-d)/2$, where r is the radius of the disc. Since the boundary of an open square centred at $\mathbf{y}$ can be at most at a distance $\sqrt{2}s$ from $\mathbf{y}$ and the edge of the disc is at least $2s$ from $\mathbf{y}$, the minimum distance from the boundary of the square to that of the disc is more than $(2-\sqrt{2})s$. Hence the open square lies entirely inside the open disc. Thus every point inside U_u can be enclosed by an open square and U_u can therefore be regarded as a union of open squares.

Theorem 3.4. The square and the usual metric generate equivalent topologies on $\mathbb{R}^2$.

Proof. Consider $U_1 \in \tau_1$ where τ_1 is the topology generated by the square metric ρ_1. By definition, every $x \in U_1$ is the center of some open square $S_r^1(\mathbf{x})$. We want to show that $U_1 \in \tau_u$, where τ_u is the usual topology, generated by ρ_u. Thus we need to prove that $\forall y \in S_r^1(\mathbf{x}) \; \exists$ an open sphere in the usual metric, $S_s^u(\mathbf{y})$ s.t. $S_s^u(\mathbf{y}) \subseteq S_r^1(\mathbf{x})$. We achieve this by constructing a specific $S_s^u(\mathbf{y})$. By definition
$$\rho_1(\mathbf{x}, \mathbf{y}) = \max\{|y_1 - x_1|, |y_2 - x_2|\}. \tag{3.43}$$
Suppose that $|y_1 - x_1| \geq |y_2 - x_2|$ and $y_1 > x_1$. This assumption does not reduce generality as we could reverse the order and use the same argument with y_2 and x_2. The only case missed out is $\mathbf{x} = \mathbf{y}$, in which case it is obvious that $S_r^u(\mathbf{y}) \subseteq S_r^1(\mathbf{x})$ as the unit disc lies inside the unit square. Let $y_1 = x_1 + d$. We can choose $s = (r-d)/2 > 0$ to provide a disc around $\mathbf{y}$ which is contained in the open square. That it is also contained in the square in the "2"–direction is guaranteed by the fact that $|y_2 - x_2| \leq y_1 - x_1$. This argument is illustrated in Figure 3.9(a). The reverse argument, that $U_u \in \tau_u \Rightarrow U_u \in \tau_1$ can be similarly shown by considering $\mathbf{x} \in U_u$, taking any $\mathbf{y} \in S_r^u(\mathbf{x})$ and showing that we can construct $s > 0$ s.t. $S_s^1(\mathbf{y}) \subseteq S_r^u(\mathbf{x})$. This is illustrated in Figure 3.9(b). Since $U \in \tau_1 \iff U \in \tau_u$, the topologies are equivalent. $\square$

The above theorem can be extended also to ρ_2, i.e., the topology defined by open diamonds. Further, the choice of $\mathbb{R}^2$ was purely for convenience of showing the argument diagrammatically. We could equally well use $\mathbb{R}^n$. These extensions are left as an exercise for you. It would be worth while to go through the construction of proofs rigorously to get a feel for the arguments and the methods involved. Let us also look at some other examples of equivalent topologies.

Example 1. On $\mathbb{R}$, the metrics
$$\rho(x,y) = |x - y|, \; \rho_n(x,y) = |x - y|/n, \; \rho_m(x,y) = m|x - y|, \tag{3.44}$$
where $m, n \in \mathbb{N}$ and $x, y \in \mathbb{R}$, generate equivalent topologies. (It is obvious that open intervals remain open on re-scaling.)

Example 2. On a finite set, X, every metric generates a topology equivalent to every other metric, as a singleton set $\{x\}$ is necessarily closed in a metric space. Thus every subset is also closed, as it is a union of closed sets. Now every subset is the complement of some subset. Hence every subset is also open. Hence the topology generated by the metric on a finite set is the discrete topology.

Example 3. The basis $\beta_d = \{[a,b]|a,b \in \mathbb{R}\}$ on $\mathbb{R}$ generates a topology equivalent to the discrete topology. This is easy to see as the intersection of $[a,b]$ and $[b,c]$ is $\{b\}$ and so singleton sets must be regarded as open. Similarly $\beta_r = \{(a,b]|a,b \in \mathbb{R}\}$ generates a topology equivalent to the right ray topology and $\beta_\ell = \{[a,b)|a,b \in \mathbb{R}\}$ to the left ray topology. The proof of this fact is left as an exercise for you.

3.4 Formulation with Closed Sets

In this section we treat the notions of closed sets, closure of a set and limit points. As already seen in the previous chapter, these concepts are crucial and a firm grasp of them is required before one can proceed on to the more intriguing aspects of Topology. In particular, no concept of Topology or Analysis is of greater importance than the notion of a limit point. It will become increasingly apparent, as we proceed, that the entire fabric of Topology is permeated by it.

As with more general topological spaces, a subset F of a metric space X is said to be *closed* if F' is open in the sense of metric spaces (as defined in Section 3.2). Thus if $U \subseteq X$ is open then U' is closed. Of course there can be sets like $(0,1]$ or $\mathbb{Q}$ which are neither open nor closed subsets of $\mathbb{R}$.

Recall that $x \in X$ is a *limit point* or *accumulation point of* $A \subseteq X$ if $\forall$ nbds $\eta \ni x, A \cap (\eta \smallsetminus \{x\}) \neq \phi$ and the set of all limit points of A is called the *derived set* of A, denoted by $\mathfrak{D}(A)$.

As before we state various theorems to provide characterizations. Here we start with the characterization of closed sets in terms of their closures. Remember that if A is closed then $\bar{A} = A$ in topological spaces. This carries through here.

Theorem 3.5. A subset A of a metric space X is closed iff $\bar{A} = A$.

Proof. It would be useful for you to provide the complete proof of this statement along the same lines as the proof for topological spaces. The concepts: 'interior', 'exterior' and 'boundary' or 'frontier' are also defined in the same way as for topological spaces. □

We now show that each closed sphere is a closed set.

Theorem 3.6. If (X, ρ) is a metric space, $\overline{S}_r(x_0) \subseteq X$ is a closed set.

Proof. From the above definition all we need to do is to show that its complement is open, i.e., that
$$A = \{x | x \in X, \rho(x, x_0) > r\} \tag{3.45}$$
is open. For a particular $x \in A$ let $\rho(x, x_0) = r_0$ and define $r_1 = r_0 - r > 0$ (see Figure 3.10). We can then construct the open sphere $S_{r_1}(x) \subseteq A$. Now, for any $y \in S_{r_1}(x), \rho(x,y) < r_1$. Using the triangular inequality, we see that
$$\rho(y, x_0) \geq \rho(x, x_0) - \rho(y, x) > r_0 - (r_0 - r) = r. \tag{3.46}$$
Hence $y \in A$ and so $S_{r_1}(x) \subseteq A$. □

Theorem 3.7. Each singleton $\{x_0\}$ is a closed set in a metric space X.

Proof. For the given $x_0 \in X$, consider the set
$$A = X \smallsetminus \{x_0\} = \{x | x \in X, \rho(x, x_0) > 0\}, \tag{3.47}$$
and now take any point $x \in A$. Let $\rho(x, x_0) = r$ and construct the open sphere $S_r(x)$. Consider the triangular inequality for the third point $y \in S_r(x)$. Then $\rho(x_0, y) \geq \rho(x, x_0) - \rho(y, x) = r - \rho(y, x) > 0$. Therefore $y \neq x_0$. Hence $y \in A$ and so $S_r(x) \subseteq A$. Hence A is open so its complement $\{x_0\}$ is closed as claimed. □

The importance of the notion of limit points in Analysis becomes clear from the following theorem.

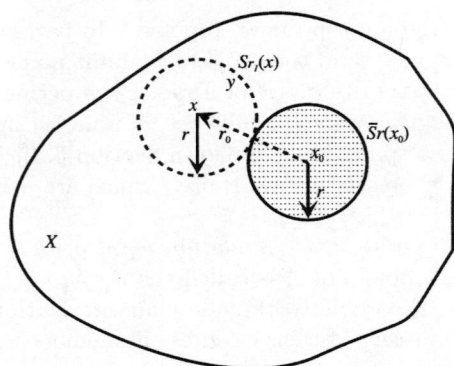

Figure 3.10: We construct the closed sphere $\overline{S}_r(x_0)$ of radius r about the point x. Now a new point $x \in X$ is taken at distance $r_0 > r$ from x_0. A new open sphere $S_{r_1}(x)$ is constructed with radius $r_1 = r_0 - r$ about x. Then any $y \in S_{r_1}(x)$ obviously lies outside $\overline{S}_r(x_0)$. Hence $A = \overline{S}'_r(x_0)$ is clearly a union of open spheres like $S_r(x_0)$.

Theorem 3.8. For a subset A of a metric space X a point $x \in X$ lies in A iff $\exists$ a sequence $\{x_n\} \subset A$ s.t. $x_n \to x$.

Proof. The proof is simple and is, therefore, left as an exercise. □

While we can have finite open sets in the context of topological spaces this is not the case for metric spaces, where the topology is induced by the metric. Using Theorem 3.7 you can prove,

Theorem 3.9. Every finite subset of a metric space is closed.

The following theorem also follows from Theorem 3.7.

Theorem 3.10. For any subset A of a metric space X and any $x \in X$, $x \in \bar{A}$ iff every nbd η of x contains infinitely many points of A.

3.5 Complete Metric Spaces

Complete metric spaces possess enough structure to enable one to establish many theorems that have wide applications in both Topology and Analysis. For instance, one theorem of topological character concerning metric spaces is a theorem relating compactness of a space to completeness. We prove it in Chapter 6. Other significant results about complete metric spaces are the Baire's Category theorem and the fixed point theorem, which can be used to prove theorems involving the existence and uniqueness of solutions to differential equations.

The motivation for the definition of a complete metric space comes from Cauchy's method of constructing the real numbers from the rationals. He showed that a sequence, in the set of rationals, may or may not converge. Nowadays it has become a hallowed tradition to use Cauchy sequence to define every concept and prove every theorem in the theory of complete metric spaces.

Recall that by a sequence $\{x_n\}$ of elements of X is meant a mapping $n \to x_n$ of the positive integers into X. A subset A of a metric space is bounded iff $diam(A) < \infty$ where

$$diam(A) = \sup_{x,y \in A} \rho(x,y). \tag{3.48}$$

In a metric space (X, ρ) a sequence $\{x_n\}$ is said to be a *Cauchy sequence* iff for every $\epsilon > 0$ there is an integer $N(\epsilon)$ s.t. $\rho(x_m, x_n) < \epsilon$ whenever $m, n > N$. Equivalently, $x_n \in S_\epsilon(x_N)$ for all $n \geq N$, or

$$\lim_{m,n \to \infty} \rho(x_m, x) = 0. \tag{3.49}$$

As an obvious consequence of this definition we have the following result in a metric space (X, ρ):
Proposition: Every Cauchy sequence in a metric space X is bounded.
Corollary 3.10.1. Every convergent sequence is bounded.

A sequence $\{x_n\}$ is said to be *convergent* if $\exists$ a point $x \in X$ s.t. $\forall \epsilon > 0 \; \exists \; N \in \mathbb{N}$ s.t.

$$\rho(x_n, x) < \epsilon, \quad \forall n \geq N. \tag{3.50}$$

Equivalently, for each open sphere $S_\epsilon(x), \exists \; N \in \mathbb{N}$ s.t. $x_n \in S_\epsilon(x) \forall n \geq N$. This means that each open sphere contains almost all the points of the sequence except some finite number of points. We usually denote this by writing $x_n \to x$ as $n \to \infty$ or $\lim_{n \to \infty} x_n = x$.

It is important to remember that the notion of a convergent sequence is not intrinsic to the sequence itself, but also depends on the structure of the space in which it lies. More precisely, a convergent sequence is not convergent on its own, but it must converge to some point in the given space.

Theorem 3.11. Let (X, ρ) be a metric space. Every convergent sequence, $\{x_n\} \in X$, is a Cauchy sequence.

Proof. Suppose $\{x_n\}$ converges to x. Let $\epsilon > 0$. Then $\exists \; N \in \mathbb{N}$ s.t. $\rho(x_n, x) < \frac{\epsilon}{2}, \forall n \geq N$. This implies that

$$\rho(x_m, x_n) \leq \rho(x_m, x) + \rho(x_n, x) < \frac{\epsilon}{2} + \frac{\epsilon}{2} = \epsilon, \tag{3.51}$$

$\forall m, n \geq N$ so that $\{x_n\}$ is a Cauchy sequence. $\square$

It is to be noted that the converse of Theorem 3.11 is not always true. In some metric spaces there are Cauchy sequences which do not converge. For instance, consider the subspace $X = (0, 1]$ of the real line $\mathbb{R}$. The sequence defined by $x_n = \frac{1}{n}$ is a Cauchy sequence in X, but it is not convergent in X, since the point 0, which this sequence wants to converge to, does not belong to X. This problem, that Cauchy sequences in a space may not be convergent sequences in it, is what leads us to define complete metric spaces.

A metric space (X, ρ) is said to be *complete* if any Cauchy sequence $\{x_n\}$ converges in X, i.e., $\exists \; x \in X$ s.t. $x_n \to x$.

Any $(\mathbb{R}^n, \tau_u)$, where $n \in \mathbb{N}$, is a complete metric space (as will be shown later). There is also no dearth of non-complete metric spaces.

Example 1. Consider $X = [0, 1) \subseteq \mathbb{R}$. Thus $\{x_n\} = \{1 - \frac{1}{n}\}$ is a Cauchy sequence that does not converge in X, as it converges to 1.

Example 2. $\mathbb{Q} \subset \mathbb{R}$ is not a complete metric space. To see this fact construct a sequence $\{x_n\}$ of rational numbers such that $x_n \to \sqrt{2}, e, \pi$ etc.

3.6 Characterization of Completeness

To verify that a space, like $\mathbb{R}^n$, is complete it is useful to have some criteria which must be fulfilled. These are given by the following theorems.

Theorem 3.12. A metric space (X, ρ) is complete iff every Cauchy sequence in X has a convergent subsequence.

Proof. Necessity is obvious. For, if X is a complete metric space and if $\{x_n\}$ is a Cauchy sequence in X, then it has a convergent subsequence, namely $\{x_n\}$ itself.

Conversely, let $\{x_n\}$ be a Cauchy sequence in X. We prove that if $\{x_n\}$ has a convergent subsequence $\{x_{n_k}\}$ that converges to a point x of X, then the entire sequence $\{x_n\}$ must also converge to x. Given $\epsilon > 0$, we choose a sufficiently large N s.t. $\rho(x_m, x_n) < \frac{\epsilon}{2}$ $\forall m, n \geq N$. We can do so because $\{x_n\}$ is a Cauchy sequence. Now choose an integer k large enough so that $n_k \geq N$ and $\rho(x_{n_k}, x) < \frac{\epsilon}{2}$. (This is possible since by hypothesis $\{x_{n_k}\}$ is a convergent subsequence.) Therefore,

$$\rho(x_n, x) \leq \rho(x_n, x_{n_k}) + \rho(x_{n_k}, x) < \frac{\epsilon}{2} + \frac{\epsilon}{2} = \epsilon, \; \forall n, n_k \geq N. \tag{3.52}$$

Hence $\{x_n\}$ converges to x and thus X is a complete metric space. $\square$

Here is another important criterion for completeness, usually called *Cantor's Intersection theorem*.

Theorem 3.13. A metric space (X, ρ) is complete iff for every sequence of closed spheres $\{\overline{S}_n\}$, with $\overline{S}_{n+1} \subset \overline{S}_n$, $n = 1, 2, \cdots$ and $\lim_{n \to \infty} r_n = 0$, where r_n is the radius of $\overline{S}_n$, the intersection $\bigcap_{n=1}^{\infty} \overline{S}_n$ consists of exactly one point.

Proof. (Necessity) Assume that X is a complete metric space and let x_n be the center of the closed sphere $\overline{S}_n$ with radius r_n. Since $\overline{S}_{n+1} \subset \overline{S}_n$, for each natural number p we have $\rho(x_{n+p}, x_n) \leq r_n$. It follows that the sequence $\{x_n\}$ of the centres of spheres $\overline{S}_n$ is a Cauchy sequence, which must converge in X because X is, by assumption, complete. Let $\lim_{n \to \infty} x_n = x$; then $x \in \bigcap_{n=1}^{\infty} \overline{S}_n$. In fact, for each $p \in \mathbb{N}, x_{n+p} \in \overline{S}_n$; this means that almost all the points of $\{x_n\}$ are contained in $\overline{S}_n$ except possibly the points $x_1, x_2, \cdots, x_{n-1}$. Therefore, x is a limit point of $\overline{S}_n$ and since $\overline{S}_n$ is closed, $x \in \overline{S}_n$ for each n. Thus, $x \in \bigcap_{n=1}^{\infty} \overline{S}_n$. The uniqueness follows immediately from the assumption that $\lim_{n \to \infty} r_n = 0$.

(Sufficiency) Let $\{x_n\}$ be a Cauchy sequence in X. We show that it has a convergent subsequence $\{x_{n_k}\}$; this will establish the completeness of X by virtue of Theorem 3.12. Since $\{x_n\}$ is a Cauchy sequence, we can choose a point x_{n_1} from the sequence $\{x_n\}$ so that $\rho(x_n, x_{n_1}) < \frac{1}{2}, \forall n \geq n_1$. Construct a closed sphere $\overline{S}_1$ with centre at x_{n_1} and radius 1. Next choose x_{n_2} from $\{x_n\}$ so that $n_2 > n_1$ and $\rho(x_n, x_{n_2}) < \frac{1}{2^2}, \forall n \geq n_2$. Denote by $\overline{S}_2$ the closed sphere with center at x_{n_2} and radius $\frac{1}{2^2}$. Analogously, we choose the point $x_{n_{k+1}}$ from $\{x_n\}$ so that $n_{k+1} > n_k$ and $\rho(x_n, x_{n_{k+1}}) < \frac{1}{2^{k+1}}, \forall n \geq n_{k+1}$. Denote by $\overline{S}_k$ the closed sphere with center at $x_{n_{k+1}}$ and radius

$\frac{1}{2^k}$. Thus we have a sequence of closed spheres $\{\overline{S}_k\}$ s.t. $\overline{S}_{k+1} \subset \overline{S}_k$ whose radii tend to zero. By hypothesis, $\exists!$ point $x \in X$ s.t. $x \in \bigcap_{k=1}^{\infty} \overline{S}_k$. Obviously, by our construction x is the limit of the sequence $\{x_{n_k}\}$ i.e., $\lim_{n_k \to \infty} x_{n_k} = x$. Hence by the preceding theorem $\{x_n\}$ converges to x and thus X is a complete metric space. □

The next theorem ensures the completeness of many metric spaces which arise as subspaces of complete metric spaces.

Theorem 3.14. A subspace Y of a complete metric space X is complete iff it is closed in X.

Proof. Let Y be a complete subspace of X. We show that it is closed i.e., $\overline{Y} = Y$. Suppose y is an arbitrary limit point of Y. For each positive integer, n, the open sphere $S_{\frac{1}{n}}(y)$, by the definition of the limit point, contains a point y_n (say) in Y, so that $\rho(y_n, y) < \frac{1}{n} \to 0$ as $n \to \infty$.
Since
$$\rho(y_m, y_n) \leq \rho(y_m, y) + \rho(y_n, y) < \frac{1}{n} + \frac{1}{m} \to 0 \text{ as } n, m \to \infty, \tag{3.53}$$
$\{y_n\}$ is a Cauchy sequence in Y which must converge in Y by virtue of its completeness. Hence $y \in Y$. Thus $\overline{Y} = Y$.

To prove the converse, we assume that Y is a closed subset of a complete space X. We have to show that it is complete. Let $\{y_n\}$ be any Cauchy sequence in Y, then obviously it is also a Cauchy sequence in X. Since, by hypothesis, X is complete, $\{y_n\}$ must converge to some point x of X. There are two possibilities: either $\{y_n\}$ has only a finite number of distinct points or $\{y_n\}$ has infinitely many distinct points. If $\{y_n\}$ has finitely many distinct points, then x is that point infinitely repeated and thus $x \in Y$. Otherwise, since $y_n \to x$ all y_n's, from some place on, must lie in $S_\epsilon(x)$, therefore x is a limit point of the set $\{y_1, y_2, \cdots, y_n, \cdots\} \subseteq Y$ and so x is a limit point of Y. Since Y is closed, $x \in Y$. This means that every Cauchy sequence in Y converges in it and hence it is complete. □

We now present examples of complete metric space.

Example 1. Consider $(\mathbb{R}, \tau_d)$. A sequence $\{x_n\}$ in $\mathbb{R}$ is a Cauchy sequence iff the points of the sequence $\{x_1, x_2, \cdots, x_p, x_p, \cdots\}$, start repeating after some p. Indeed, if $\{x_n\}$ is a Cauchy sequence, then for every $\epsilon > 0 \ \exists \ N$ s.t. $\rho(x_m, x_n) < \epsilon, \forall m, n \geq N$. Choose $\epsilon = \frac{1}{2}$, then $\rho(x_m, x_n) < \epsilon$ whenever $x_m = x_n, \forall m, n \geq N$. Therefore, obviously $\{x_n\}$ converges in $\mathbb{R}$ and thus $(\mathbb{R}, \tau_d)$ is complete.

Example 2. $(\mathbb{R}, \rho)$ is a complete space, with $\rho(x, y) = |x - y|$. The completeness of $\mathbb{R}$ follows from the well-known results in Analysis specified below (see for instance, [Bartle and Sherbert, 2011]).

Theorem 3.15 (Monotone Convergence Theorem). If $\{x_n\}$ is a sequence in $\mathbb{R}$ which is monotonically non-decreasing i.e., $x_1 \leq x_2 \leq \cdots \leq x_n \leq x_{n+1} \leq \cdots$ then $\{x_n\}$ converges in $\mathbb{R}$ iff it is bounded, in which case $\lim_{n \to \infty} x_n = \sup_n \{x_n\}$. Similarly, if a sequence is monotonically decreasing, i.e., $x_1 \geq x_2 \geq \cdots \geq x_n \geq x_{n+1} \geq \cdots$ then $\{x_n\}$ converges iff it is bounded, in which case $\lim_{n \to \infty} x_n = \inf_n \{x_n\}$.

This theorem is extraordinarily useful and important, but it has the drawback that it applies only to sequences which are monotonic. It behooves us, therefore, to find a condition which will imply convergence in $\mathbb{R}$ without using the monotonicity. Recall that every Cauchy sequence in $\mathbb{R}$ is bounded. The condition required is provided by the following theorem (stated without proof).

Theorem 3.16 (Bolzano-Weierstrass Theorem). A bounded sequence in $\mathbb{R}$ has a convergent subsequence.

Now assume that $\{x_n\}$ is any Cauchy sequence, then it is bounded. Therefore, by the Bolzano-Weierstrass theorem theorem it has a convergent subsequence. Hence, by virtue of Theorem 3.12 the entire sequence $\{x_n\}$ and thus $\mathbb{R}$ is complete. This fact can be reformulated as the *Cauchy Convergence Criterion*: A sequence $\{x_n\}$ in $\mathbb{R}$ is convergent iff it is a Cauchy sequence.

Example 3. $(\mathbb{R}^n, \rho)$ is complete with

$$\rho(x,y) = \sqrt{\sum_{k=1}^{n}(y_k - x_k)^2}. \tag{3.54}$$

Let $\{\mathbf{x}^{(p)}\}$ be any Cauchy sequence in $\mathbb{R}^n$, where $\mathbf{x}^{(p)} = \left(x_1^{(p)}, x_2^{(p)}, \cdots, x_n^{(p)}\right)$, $p = 1, 2, \cdots$. Then for each $\epsilon > 0, \exists$ a natural number $N(\epsilon) = N$ s.t.

$$\sum_{k=1}^{n}\left(x_k^{(p)} - x_k^{(q)}\right)^2 < \epsilon^2, \quad \forall \ p, q \geq N. \tag{3.55}$$

It follows that, for each k, the sequence of numbers $\left\{x_k^{(q)}\right\}$ is a Cauchy sequence in $\mathbb{R}$, since $\left|x_k^{(p)} - x_k^{(q)}\right| < \epsilon, \forall p, q \geq N$. Since $\mathbb{R}$ is complete, let $x_k = \lim_{p \to \infty} x_k^{(p)}, k = 1, 2, \cdots, n$, and let $x = (x_1, \cdots, x_n)$. Then obviously $\lim_{p \to \infty} x^{(p)} = x$. Hence every Cauchy sequence in $\mathbb{R}^n$ is complete.

Similar arguments yield that $(\mathbb{R}^n, \rho)$ is also complete with

$$\rho(x,y) = \max_{1 \leq k \leq n} |y_k - x_k|. \tag{3.56}$$

Example 4. The space

$$\ell_2 = \left\{\mathbf{x} = (x_1, x_2, \cdots, x_n, \cdots) \mid \sum_{k=1}^{\infty} x_k^2 < \infty, \ x_k \in \mathbb{R}\right\} \tag{3.57}$$

is complete. Let $\{x^{(n)}\}$ be a Cauchy sequence in ℓ_2. Then by definition, for every $\epsilon > 0 \ \exists \ N \in \mathbb{N}$ s.t. $N = N(\epsilon)$ and

$$\rho(x^{(n)}, x^{(m)}) \leq \sqrt{\sum_{k=1}^{\infty}\left(x_k^{(n)} - x_k^{(m)}\right)^2} < \epsilon \ \forall n, m \geq N(\epsilon). \tag{3.58}$$

Here

$$\mathbf{x}^{(n)} = \left(x_1^{(n)}, x_2^{(n)}, \cdots, x_k^{(n)}, \cdots\right). \tag{3.59}$$

It follows from Inequality (3.58) that, for each k, $\{x_k^{(n)}\}$ is a Cauchy sequence in $\mathbb{R}$, and since $\mathbb{R}$ is complete, $\lim_{n\to\infty} x_k^{(n)} = x_k$ exists. Denoting $(x_1, x_2, \cdots, x_k, \cdots)$ by $\mathbf{x}$ we must show that

$$\sum_{k=1}^{\infty} x_k^2 < \infty, \quad \text{i.e., } \mathbf{x} \in \ell_2; \tag{3.60}$$

$$\lim_{n\to\infty} x^{(n)} = x. \tag{3.61}$$

From Inequality (3.58), it follows that for each fixed M

$$\sum_{k=1}^{M} \left(x_k^{(n)} - x_k^{(m)}\right)^2 < \epsilon^2. \tag{3.62}$$

In this sum, now there are only finitely many terms of the sequence. Fixing n and letting $m \to \infty$, we obtain

$$\sum_{k=1}^{M} \left(x_k^{(n)} - x_k\right)^2 \leq \epsilon^2, \tag{3.63}$$

which holds for every M. Now letting $M \to \infty$, we obtain

$$\sum_{k=1}^{\infty} \left(x_k^{(n)} - x_k\right)^2 < \epsilon^2, \quad \forall n \geq N. \tag{3.64}$$

From Inequality (3.64) it follows that the sequence $\{x^{(n)} - x\} \in \ell_2$. From the convergence of the series $\sum_{k=1}^{\infty} \left(x_k^{(n)}\right)^2$ and $\sum_{k=1}^{\infty} \left(x_k^{(n)} - x_k\right)^2$ follows the convergence of the series $\sum_{k=1}^{\infty} x_k^2$ (by virtue of the elementary Inequality $(a+b)^2 \leq 2(a^2 + b^2)$). Thus inequality (3.60) is proved. Next, since ϵ is chosen arbitrarily, the inequality (3.64) means that

$$\lim_{n\to\infty} \rho(x^{(n)}, x) = \lim_{n\to\infty} \sqrt{\sum_{k=1}^{\infty} \left(x_k^{(n)} - x_k\right)^2} = 0, \tag{3.65}$$

and therefore Eq. (3.61) is also proved. Hence ℓ_2 is a complete metric space.

Exercise: Show that the space ℓ_p, for $1 \leq p < \infty$, of all sequences $x = (x_1, x_2, \cdots, x_n, \cdots)$ satisfying $\sum_{k=1}^{\infty} |x_k|^p < \infty$ with the metric

$$\rho(x, y) = \left[\sum_{k=1}^{\infty} |x_k - y_k|^p\right]^{\frac{1}{p}}, \tag{3.66}$$

is a complete metric space.

Example 5. Consider the space ℓ_∞ of all bounded sequences, i.e., all $\mathbf{x} = (x_1, x_2, \cdots, x_n, \cdots)$ for which $\sup_{1 \leq k < \infty} |x_n| < \infty$, with the metric

$$\rho_\infty(x, y) = \sup_{1 \leq n < \infty} |x_n - y_n|. \tag{3.67}$$

We show that ℓ_∞ is a complete space. For this purpose let $\{x^{(n)}\}$ be any Cauchy sequence in ℓ_∞, where $\mathbf{x}^{(n)} = \left(x_1^{(n)}, x_2^{(n)}, \cdots, x_k^{(n)}, \cdots\right)$. Then for each $\epsilon > 0$, there is an N s.t.

$$\sup_{1 \leq n < \infty} \left|x_k^{(n)} - x_k^{(m)}\right| < \epsilon, \ \forall n, m \geq N. \tag{3.68}$$

It follows from here that, for each k,

$$\left|x_k^{(n)} - x_k^{(m)}\right| < \epsilon, \ \forall n, m \geq N, \tag{3.69}$$

and hence $\{x_k^{(n)}\}$ is a Cauchy sequence in $\mathbb{R}$. Since $\mathbb{R}$ is complete, $\lim_{n \to \infty} x_k^{(n)} = x_k$ (say) exists. Now fixing n and letting $m \to \infty$ in Inequality (3.68), we obtain

$$\left|x_k^{(n)} - x_k\right| \leq \epsilon, \ \forall n \geq N, \tag{3.70}$$

for each k. Therefore

$$\sup_{1 \leq k < \infty} \left|x_k^{(n)} - x_k\right| \leq \epsilon, \ \forall n \geq N, \tag{3.71}$$

from which it follows that $x^{(n)} \to x$. Since

$$|x_k| \leq \left|x_k^{(n)}\right| + \left|x_k - x_k^{(n)}\right|, \tag{3.72}$$

we have

$$\sup_{1 \leq k < \infty} \left|x_k^{(n)} - x_k\right| \leq M + \epsilon, \tag{3.73}$$

which implies that $x \in \ell_\infty$. Thus ℓ_∞ is a complete metric space.

Example 6. We can show that the space $C([a, b])$ is complete with respect to the metric

$$\rho(f, g) = \sup_{1 \leq t \leq b} |f(t) - g(t)|. \tag{3.74}$$

To proceed with the proof, let $\{f_n\}$ be a Cauchy sequence in $C([a, b])$. Then for every $\epsilon > 0$ there is an N s.t. $m, n > N$ implies

$$\rho(f_n, f_m) < \epsilon. \tag{3.75}$$

But this means that

$$\sup_{1 \leq t \leq b} |f_n(t) - g_m(t)| < \epsilon. \tag{3.76}$$

Therefore, for every $t \in [a, b]$

$$|f_n(t) - g_m(t)| < \epsilon, \ \forall n, m \geq N, \tag{3.77}$$

i.e., $\{f_n(t)\}$ is a Cauchy sequence in $\mathbb{R}$. By the completeness of $\mathbb{R}$, $\{f_n(t)\}$ converges to a real number $f(t)$, so that we have determined a function f on $[a, b]$ defined by $f(t) = \lim_{n\to\infty} f_n(t)$, for every $t \in [a, b]$. (This convergence is pointwise.)

Note that f is a continuous function on $[a, b]$. Fixing n and letting $m \to \infty$ in the above inequality, we obtain

$$|f_n(t) - f(t)| \leq \epsilon \tag{3.78}$$

for every $t \in [a, b]$ and $\forall n \geq N$. Therefore,

$$\sup_{a \leq t \leq b} |f_n(t) - f(t)| \leq \epsilon, \ \forall n \geq N. \tag{3.79}$$

This implies that $\{f_n\}$ converges uniformly to f and $\lim_{n\to\infty} \rho(f_n, f) = 0$. Thus $C([a, b])$ is complete.

Example 7. Let $C([a, b])$ be the space of all continuous functions defined on $[a, b]$ with respect to the metric given by

$$\rho(f, g) = \int_a^b |f(t) - g(t)| \, dt. \tag{3.80}$$

(As an exercise verify that ρ is, in fact, a metric on $C([a, b])$.) We show that it is *not* complete. Let $c \in (a, b)$ and for every positive integer, n, so large that $a < c - 1/n$ define $f_n(t)$ as

$$f_n(t) = \begin{cases} 0 & \text{if } a \leq t \leq c - \frac{1}{n}, \\ nt - nc + 1 & \text{if } c - \frac{1}{n} \leq t \leq c, \\ 1 & \text{if } c \leq t \leq b. \end{cases} \tag{3.81}$$

Then $\rho(f_n, f_m) \leq \frac{1}{n} + \frac{1}{m} \to 0$ as $n, m \to \infty$. Therefore $\{f_n\}$ is a Cauchy sequence in $C([a, b])$. Now, let $f \in C([a, b])$. Thus

$$\rho(f_n, f) = \int_a^{c-\frac{1}{n}} |f(t)| \, dt + \int_{c-\frac{1}{n}}^c |f_n(t) - f(t)| \, dt + \int_c^b |1 - f(t)| \, dt \tag{3.82}$$

and $\lim_{n\to\infty} \rho(f_n, f) = 0$. This implies that $f(t) = 0$, $t \in [a, c)$ and $f(t) = 1, t \in (c, b]$. Since it is impossible for a continuous function to have this property, $\{f_n\}$ does not have a limit. Hence $C([a, b])$ is not complete with respect to this metric.

3.7 Further Theorems Related to Completeness

To be able to state the further theorems we need to introduce some new concepts which require new definitions. Let (X, ρ) be a metric space. A subset A of X is said to be *nowhere-dense* in X iff $\bar{A}$ contains no open subset of X, i.e., $int(\bar{A}) = \phi$. To clarify this concept we present a couple of examples.

Example 1. $\mathbb{Z}$ is nowhere-dense in $(\mathbb{R}, \rho_u)$, since $int(\overline{\mathbb{Z}}) = int(\mathbb{Z}) = \phi$. Similarly, $\mathbb{N}$ is nowhere-dense in $(\mathbb{R}, \rho_u)$. However, $\mathbb{Q}$ is not nowhere-dense in $(\mathbb{R}, \rho_u)$ since $int(\overline{\mathbb{Q}}) = int(\mathbb{R}) = \mathbb{R} \neq \phi$.

Example 2. In $(\mathbb{R}^2, \rho_u)$ consider the straight line passing through the origin

$$A = \mathbb{R} \times \{0\} = \{(x, 0) : x \in \mathbb{R}\}. \tag{3.83}$$

Then $int(\bar{A}) = int(A) = \phi$, and thus A is nowhere-dense in $(\mathbb{R}^2, \rho_u)$. The same argument applies to any line

$$L = \{(x, y) | y = mx + c\}. \tag{3.84}$$

We now come to the important definition for this section.

A subset A of a metric space X is said to be a set of the *first category* in X iff A can be written as a countable union of nowhere dense sets in X, i.e., $A = \bigcup_{k=1}^{\infty} A_k$, where each A_k is nowhere-dense in X. The subset A is said to be of the *second category* if A is not of the first category. We again provide an example.

Example 3. The set $\mathbb{Q}$ is of the first category in $(\mathbb{R}, \tau_d)$, since $\mathbb{Q} = \bigcup_{x \in \mathbb{Q}} \{x\}$. Note that each $\{x\}$ is nowhere-dense in $\mathbb{R}$, since $int(\overline{\{x\}}) = int(\{x\}) = \phi$. Similarly, $\mathbb{Z}$ and $\mathbb{N}$ are of the first category. However, the set of irrationals, $\mathbb{Q}'$, is of the second category. Since, if the irrationals could be written as a countable union of nowhere-dense sets, with the addition of the individual rational points as nowhere-dense subsets, $\mathbb{R}$ itself could be represented as a countable union of nowhere-dense subsets and thus would be of the first category in itself. However, this contradicts Baire's Category Theorem which asserts that every complete metric space is of the second category. Before going on to that theorem we need to present another theorem.

Theorem 3.17. Let (X, ρ) be a complete metric space and $A_1, A_2, \cdots, A_n, \cdots$ be a sequence of open dense subsets of X. Then $A = \bigcap_{n=1}^{\infty} A_n$ is dense in X.

Proof. It is sufficient to show that $U \cap A \neq \phi \forall\, U \in \tau_\rho$. Since A_1 is open and dense in X, $U \cap A_1$ must be non-empty and open. Hence we can find a positive number $r_1 < 1$ and a point $x_1 \in X$ s.t. the closed sphere $\overline{S_{r_1}(x_1)} \subseteq U \cap A_1$. The open set $S_{r_1}(x_1)$ has non-empty intersection with A_2, and consequently there is a point x_2 and a positive number $r_2 < \frac{1}{2}$ s.t. $\overline{S_{r_2}(x_2)} \subset A_2 \cap S_{r_1}(x_1)$. Similarly, $S_{r_2}(x_2) \cap A_3$ is open and non-empty and therefore there is an $r_3 < \frac{1}{3}$ etc. Repeated application of this procedure leads to a decreasing sequence of closed spheres with radii converging to 0. From Theorem 3.13, there is a point x common to each $\overline{S_{r_n}(x_n)}$. Hence x is contained in both U and each $A_n, n \in \mathbb{N}$, i.e., $x \in U \cap A$. This completes the proof. $\square$

We now come to Baire's Category Theorem.

Theorem 3.18. Every complete metric space (X, ρ) is of the second category in itself.

Proof. Let us assume to the contrary that the complete metric space (X, ρ) is of the first category. Then X can be written as a countable union of nowhere-dense sets $A_1, A_2, \cdots$, i.e., $X = \bigcup_{n=1}^{\infty} A_n$. Since each A_n is nowhere-dense in X, $\bar{A}_n'$ is dense in X. Also, by definition, $\bar{A}_n'$ is open for each

$n \in \mathbb{N}$. Furthermore, by virtue of the preceding theorem $\bigcap_{n=1}^{\infty} \bar{A}'_n$ is non-empty and dense in X. But

$$\bigcap_{n=1}^{\infty} \bar{A}'_n = \bigcap_{n=1}^{\infty} (X \smallsetminus \bar{A}'_n) = X \smallsetminus \left(\bigcup_{n=1}^{\infty} \bar{A}'_n \right) = \phi, \tag{3.85}$$

(as each $\bar{A}'_n$ is dense in X) which is an obvious contradiction. Hence our assumption is wrong and therefore X is of the second category. $\square$

Corollary 3.18.1. *Every countable complete metric space has an isolated point.*

As an application of the preceding theorem, one can prove the existence of continuous functions on $[0, 1]$ which are nowhere differentiable. For this purpose, consider the metric space $C_0([0, 1])$ of all continuous functions, f, satisfying the condition $f(0) = f(1)$, with the metric defined as follows:

$$\rho(f, g) = \sup\{|f(x) - g(x)| : x \in [0, 1]\}. \tag{3.86}$$

Thus $C_0([0, 1])$ is a complete metric space. So the set of functions in $C_0([0, 1])$, which are somewhere differentiable, form a subset of the first category (verify it!). Since $C_0([0, 1])$ is complete, it is of the second category and hence there exist continuous functions which are nowhere differentiable. In fact, we have the following theorem.

Theorem 3.19. *Let $f : [0, 1] \to \mathbb{R}$ be a continuous function. Given $\epsilon > 0$, there is a function $g : [0, 1] \to \mathbb{R}$ with $|f(x) - g(x)| < \epsilon \ \forall \ x$, s.t. g is continuous and nowhere differentiable.*

(The proof is very lengthy and is therefore omitted.) An example of such a function is shown in Figure 3.11 and Figure 3.12, and algebraically as,

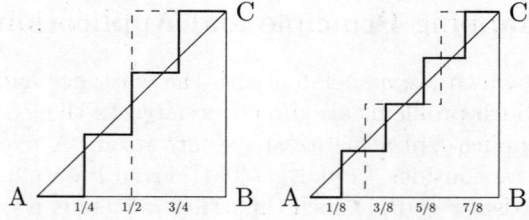

Figure 3.11: For an isosceles right triangle with unit base, ABC, one can make successive approximations to the hypotenuse AC by horizontal and vertical lines starting at A and ending at C, as shown above. The larger the number of steps, the lesser the maximum distance between the approximating lines and AC, but the length of the line remains 2. In the limit of infinitely many steps, the approximating line is visually indistinguishable from AC, whose length is $\sqrt{2}$. However it is very different, as it is that "function" that is everywhere continuous and nowhere differentiable, since it continuously changes direction. Had that not been the case, we would have "proved" $2 = \sqrt{2}$.

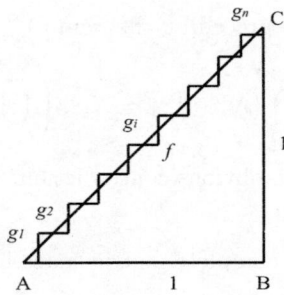

Figure 3.12: Consider the triangle with unit sides AB and BC. Let f be the function whose graph is AC. Clearly its length $\int_A^C ds = \sqrt{2}$.

$$\begin{aligned}
g_n &= 0 \times 2^{1-n} && [0 \leq x < 1/2^{n-1}) \\
&= 1 \times 2^{1-n} && [1/2^{n-1} \leq x < 2/2^{n-1}) \\
&= 2 \times 2^{1-n} && [2/2^{n-1} \leq x < 3/2^{n-1}) \\
&\vdots \\
&= 2^{n-1} \times 2^{1-n} && [(2^{(n-1)} - 1)/2^n \leq x \leq 2^{n-1}/2^{n-1}],
\end{aligned}$$

with $n \geq 3$. Clearly, as $n \to \infty$, the graph of g_n tends to the graph of f. However, the function $g = \lim_{n \to \infty} g_n \neq f$. For example, the length of g_n is 2 $\forall n$. Hence the length of g (which is $2 \times \lim_{n \to \infty} g_n$) is an example of a function that is everywhere continuous, is arbitrarily close to a differentiable function but is nowhere differentiable!

3.8 Contraction Mapping Principle and Applications

There are many problems which are associated with the existence and uniqueness of solutions to differential equations. Such problems are directly related to the existence and uniqueness of a fixed point under a mapping f of a metric space into itself. A fixed point of a mapping is most easily understood if we consider $f : [0, 1] \to [0, 1]$. From Figure 3.13 it is obvious that if f is continuous then there exists $\overline{x} \in (0, 1)$ such that $f(\overline{x}) = \overline{x}$. This is called a fixed point. (The choice of a unit interval is not crucial as a re-scaling can be introduced, but in that case we would no longer be able to state the condition that $f(\overline{x}) = \overline{x}$.) Regarding an iterative procedure to solve a differential equation as a function we can re-scale the domain and range of the solution to be the unit interval. Thus convergence of the procedure is equivalent to the existence of a fixed point. There are many different ways to determine the existence and uniqueness of fixed points. One such simple way is given below.

The 're-scaling' of the interval can be due to a shift of the starting point and/or a simple change of scale. The essence of the proof of the fixed point was due to the starting point being the same. In Figure 3.14 we see that $f : [0, a] \xrightarrow{into} [0, b]$ will also, by the same argument have a 'scaled fixed point', i.e., $f(\overline{x}) = b\overline{x}/a$. However, it is not necessary, in this case that a fixed point exists in the original sense. In Figure 3.15 we see that if $b \leq a$ it is guaranteed *without scaling* but if $b > a$ it is not. As such we define $f : (X, \rho) \xrightarrow{into} (X, \rho)$ to be a *contraction mapping* if $\exists\, \alpha > 1$ such that

$$\rho(f(x), f(y)) \leq \alpha \rho(x, y)\ \forall x, y \in X. \tag{3.87}$$

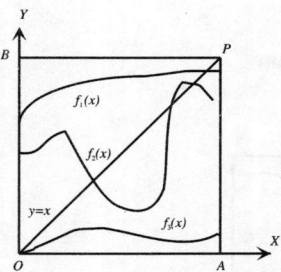

Figure 3.13: Consider the function f from OA into OB, both of unit length. The graph of this function would cover the complete domain OA but not necessarily the complete set OB. Clearly, if it is continuous, and the function starts on OB and ends on AP it must intersect OP at least at one point. It can intersect at various points and the intersection could even be at O or P.

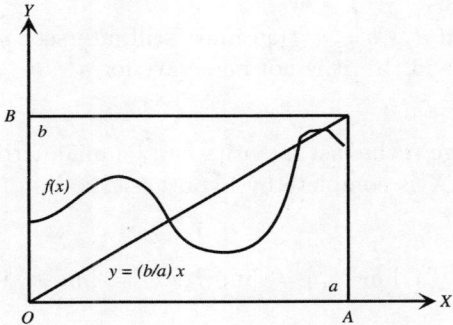

Figure 3.14: The same argument applies for $f : [o,a] \overset{into}{\longrightarrow} [o,b]$ as $y = f(x)$ must intersect $y = (b/a)\,x$ to go from a point on OB to a point on AP. The intersection point again need not be unique.

Every contraction is a continuous function. In fact, if $x_n \to x$, then by virtue of Eq. (3.87) $f(x_n) \to f(x)$. We thus have the Contraction Mapping Principle expressed by the following theorem.

Theorem 3.20. Let (X, ρ) be a complete metric space and $f : X \to X$ be a contraction. Then f has a unique fixed point.

Proof. Let x_0 be an arbitrary point of X. Put $x_1 = f(x_0)$, $x_2 = f(x_1) = f^2(x_0)$. In general, $x_n = f(x_{n-1}) = f^n(x_0)$. We show that $\{x_n\}$ is a Cauchy sequence in X. For the sake of definiteness assume that $m \leq n$. Then

$$\begin{aligned}
\rho(x_n, x_m) &= \rho(f^n(x_0), f^m(x_0)) \leq \alpha^n \rho(x_0, x_{n-m}) \\
&\leq \alpha^n \{\rho(x_0, x_1) + \rho(x_1, x_2) + \cdots + \rho(x_{n-m-1}, x_{n-m})\} \\
&\leq \alpha^n \rho(x_0, x_1) \{1 + \alpha + \alpha^2 + \cdots + \alpha^{n-m-1}\} \\
&\leq \alpha^n \rho(x_0, x_1) \frac{1}{1-\alpha} < \epsilon.
\end{aligned} \qquad (3.88)$$

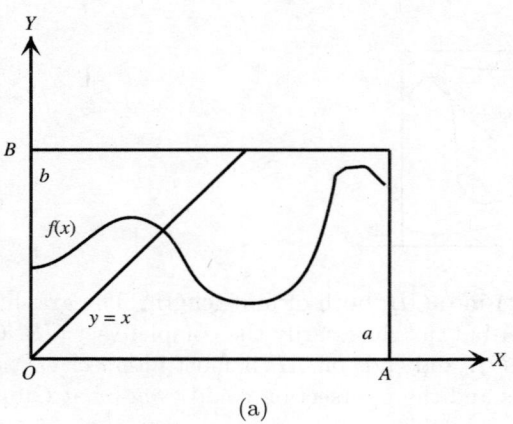
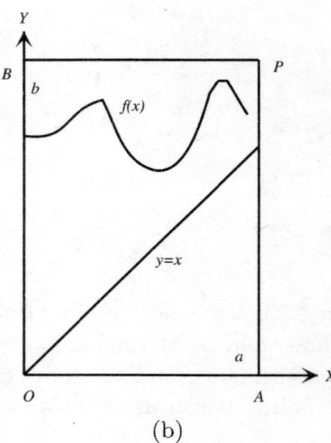

Figure 3.15: As seen in (a), if $a > b, y = f(x)$ must still intersect $y = x$ to go from OB to AP, but as is clear by the example in (b), it is not necessary for $a < b$.

(As $\alpha < 1$, for sufficiently large n the last quantity can be made arbitrarily small.) Hence, $\{x_n\}$ is a Cauchy sequence. Since X is complete by hypothesis, so $\lim_{n \to \infty} x_n = x$ (say) exists in X. Then, by the continuity of f,

$$f(x) = f\left(\lim_{n \to \infty} x_n\right) = \lim_{n \to \infty} f(x_n) = \lim_{n \to \infty} x_{n+1} = x. \tag{3.89}$$

Thus the existence of the fixed point x is proved. Now we prove its uniqueness. Suppose y is another fixed point (distinct from x). Then $\rho(x,y) \neq 0$. Hence

$$\rho(x,y) = \rho(f(x), f(y)) \leq \alpha\rho(x,y) < \rho(x,y), \tag{3.90}$$

which yields a contradiction. This completes the proof of the theorem. □

We now present some applications of contraction mappings to illustrate their significance.

The contraction mapping principle is often used to prove the existence of a solution of a nonlinear (say ordinary) differential equation. The problem is that such equations may, or may not, have any solution. If we try to find a solution by numerical methods, we will truncate the procedure somewhere and this may lead to the spurious appearance of a solution that is not really there. For, example, the procedure may lead to the summation of a harmonic series. We know that the series diverges, but it looks more and more as if it has converged as we proceed to more and more terms. Without actually summing up the series we need to know that it can be summed up. This is typically done by finding solutions to a simpler differential equation, whose graph must always lie above any solution of the equation and that must always lie below it. If we now can introduce a parameter that can be taken to tend to zero, the two graphs will "squeeze together" to give the desired solution. This way of proving the existence of solutions has the advantage that it can give an approximation of the solution that could, in principle, be used to obtain as precise an approximation as one chooses (provided we are ready to continue to carry it sufficiently far forward). Read the following example in the light of this explanation.

Let $f : [a, b] \to [a, b]$ be a function satisfying the Lipschitz condition (which ensures that a small change in some parameter appearing in the differential equation does not lead to an uncontrollably large change in the solution):

$$|f(x_2) - f(x_1)| \leq M |x_2 - x_1|, \tag{3.91}$$

where the constant $M < 1$. Then f is a contraction mapping, and as shown above, the sequence $x_0, x_1 = f(x_0), x_2 = f^2(x_0), \cdots, x_n = f^n(x_0), \cdots$ converges to a unique fixed point $f(x) = x$. In particular, f is also a contraction mapping if its derivative $f'(x)$ exists such that $|f'(x)| \leq M < 1$.

Now consider the equation $F(x) = 0$, s.t. $F(a) < 0$, $F(b) > 0$ and $0 < M_1 \leq F'(x) \leq M_2$ on the closed interval $[a, b]$. Define a function $f(x) = x - \lambda F(x)$. We now try to find the solution of the equation $x = f(x)$, which is equivalent to the equation $F(x) = 0$ for $\lambda \neq 0$. Since

$$f'(x) = 1 - \lambda F'(x), \tag{3.92}$$

therefore

$$1 - \lambda M_2 \leq f'(x) \leq 1 - \lambda M_1. \tag{3.93}$$

As another application of the previous result, we mention the following theorem whose proof is omitted. It guarantees the existence of a solution of the initial value problem.

Theorem 3.21. If f is continuous on an open connected set $\mathcal{D} \subset \mathbb{R}^2$ and satisfies a Lipschitz condition on y in $\mathcal{D}$ i.e.,

$$|f(x, y_1) - f(x, y_2)| \leq M |y_1 - y_2|, \tag{3.94}$$

then for every $(x_0, y_0) \in \mathcal{D}$, the differential equation $y'(x) = f(x, y)$ has a unique local solution passing through (x_0, y_0).

This is the fundamental theorem for the existence of a solution of a first order differential equation and gives some idea of the importance of topological considerations for the solution of differential equations.

3.9 The Completion of a Metric Space

In this section we prove a classical theorem that every metric space can be imbedded isometrically (having the same measure of magnitude) in a complete metric space. This theorem is very useful in many ways (although we shall *not* need it in our forthcoming discussions). The importance of being able to imbed a non-complete metric space into a complete one is that many results of calculus only hold on the complete space. One can use those results on the complete space and *then* restrict ourselves to the non-complete space and see what the results obtained will now imply. The enlarged space is called the *completion* of the metric space (X, ρ).

A mapping, f, of a metric space, (X, ρ_1), into another metric space, (Y, ρ_2), is called an *isometry* if, $\forall x_1, x_2 \in X$,

$$\rho_2 \left(f(x_1), f(x_2) \right) = \rho_1(x_1, x_2). \tag{3.95}$$

Such an isometry is said to be an *isometric imbedding* of X into Y.

Example 1. Take $(X, \rho_x) = (Y, \rho_y) = (\mathbb{R}, \rho_u)$. Define f by $f(x) = x + 3$. Then f is obviously an isometry.

Example 2. Consider the projection from the plane, $f: \mathbb{R}^2 \to \mathbb{R}$ given by $f(x_1, x_2) = x_1$. Then f is obviously not an isometry.

Example 3. Let $X = C([0,1])$ and $Y = \mathbb{R}$. Define $F_{x_0}(f) = f(x)$. Then for some fixed $x_0 \in [0,1]$,

$$\rho_2\left(F_{x_0}(f), F_{x_0}(g)\right) = |f(x_0) - g(x_0)|, \tag{3.96}$$

where

$$\rho(f,g) = \max\{|f(x) - g(x)| : x \in [0,1]\}. \tag{3.97}$$

Hence F_{x_0} is not an isometry.

Theorem 3.22. If (X, ρ) is any metric space, it can be imbedded, as a dense subspace, in a complete metric space $(\widetilde{X}, \widetilde{\rho})$. Two complete metric spaces in which (X, ρ) can be so imbedded are always isomorphic.

Proof. First note that $(\mathbb{R}, \rho_u)$ is complete. Let S be the set of all Cauchy sequences in X. Define an equivalence relation in S by letting $\{x_n\} \cong \{y_n\}$ if $\{\rho(x_n, y_n)\}$ converges to 0. Denote by $\widetilde{X}$ the set of all equivalence classes formed in this way. We define the metric $\widetilde{\rho}$ in $\widetilde{X}$ in the following way. If $\{x_n\}, \{y_n\} \in S$, then $\{\rho(x_n, y_n)\}$ is a Cauchy sequence of real numbers because

$$|\rho(x_n, y_n) - \rho(x_m, y_m)| \leq \rho(x_n, x_m) + \rho(y_n, y_m) < \epsilon, \ \forall \ n, m \geq N(\epsilon). \tag{3.98}$$

Thus $\{\rho(x_n, y_n)\}$ converges in $\mathbb{R}$. Let $\widetilde{\rho}$ be defined as $\lim_{n \to \infty} \rho(x_n, y_n)$. This limit is unchanged if $\{x_n\}$ and $\{y_n\}$ are replaced by equivalent sequences (verify it). Then

$$\widetilde{\rho}(\widetilde{x}, \widetilde{y}) = \lim_{n \to \infty} \rho(x_n, y_n), \tag{3.99}$$

where

$$\{x_n\} \in \widetilde{x} \in \widetilde{X}, \ \{y_n\} \in \widetilde{y} \in \widetilde{X}. \tag{3.100}$$

We only need to show that $\widetilde{\rho}$ satisfies the triangular inequality. Since in X, we have

$$\rho(x_n, z_n) \leq \rho(x_n, y_n) + \rho(y_n, z_n), \tag{3.101}$$

hence

$$\lim_{n \to \infty} \rho(x_n, z_n) \leq \lim_{n \to \infty} \rho(x_n, y_n) + \lim_{n \to \infty} \rho(y_n, z_n), \tag{3.102}$$

or

$$\widetilde{\rho}(\widetilde{x}, \widetilde{z}) \leq \widetilde{\rho}(\widetilde{x}, \widetilde{y}) + \widetilde{\rho}(\widetilde{y}, \widetilde{z}). \tag{3.103}$$

Thus $(\widetilde{X}, \widetilde{\rho})$ is a metric space. By construction it contains all sequences $(x, x, x, \cdots x, \cdots)$. That is, for each $x \in X$ there corresponds an equivalence class $\widetilde{x}$ of Cauchy sequences which converge to x. If $x = \lim_{n \to \infty} x_n$ and $y = \lim_{n \to \infty} y_n$, then

$$\rho(x, y) = \lim_{n \to \infty} \rho(x_n, y_n) = \widetilde{\rho}(\widetilde{x}, \widetilde{y}). \tag{3.104}$$

Thus the correspondence $x \to \widetilde{x}$ is an isometry. $\square$

We have earlier seen that for each $x \in X$ there corresponds an equivalence class $\widetilde{x}$ of Cauchy sequences which converge to x. This class is non-empty since the constant sequence $\{x, x, x, \cdots, x, \cdots\}$ converges to x. Thus, if $x = \lim_{n \to \infty} x_n$ and $y = \lim_{n \to \infty} y_n$, then Eq. (3.104) holds. Therefore if we associate to each $x \in X$ its equivalence class $\widetilde{x}$ of Cauchy sequences converging to x, i.e., $x \to \widetilde{x}$, we see that this mapping is an isometric imbedding of X into $\widetilde{X}$. This way we can identify X as a subspace of $\widetilde{X}$. Next, we show that X is dense (as a subspace) in $\widetilde{X}$. Let $\widetilde{x} \in \widetilde{X}$ be an arbitrary element and $\epsilon > 0$. Choose some representative Cauchy sequence. Then $\exists \, N(\epsilon)$ s.t.

$$\rho(x_m, y_n) < \epsilon, \; \forall m, n \geq N(\epsilon). \tag{3.105}$$

Then we have

$$\widetilde{\rho}(x_n, \widetilde{x}) = \lim_{m \to \infty} \rho(x_n, x_m) \leq \epsilon, \; \forall n \geq N. \tag{3.106}$$

Hence every *nbd* of $\widetilde{x}$ contains some point of X s.t. $\lim_{n \to \infty} \widetilde{\rho}(x_n, \widetilde{x}) = 0$. Thus for arbitrary $\widetilde{x} \in \widetilde{X}$ and $\epsilon > 0$, $\exists$ a point $x \in X$ s.t. $\widetilde{\rho}(\widetilde{x}, x) < \epsilon$. Thus X is dense in $\widetilde{X}$.

The space $\widetilde{X}$ is complete. In fact, suppose $\{\widetilde{x}_n\}$ is a Cauchy sequence in $\widetilde{X}$. By what we proved above, for every $\widetilde{x}_n$ $\exists$ a point $x_n \in X$ s.t.

$$\widetilde{\rho}(\widetilde{x}_n, x_n) < \frac{1}{n}, \tag{3.107}$$

and therefore $\widetilde{x} = \{x_1, \cdots, x_n, \cdots\}$ is a Cauchy sequence in X. As we have seen, $\lim_{n \to \infty} \widetilde{\rho}(x_n, \widetilde{x}) = 0$. Hence we conclude that $\lim_{n \to \infty} \widetilde{\rho}(\widetilde{x}_n, \widetilde{x}) = 0$ and thus $\widetilde{X}$ is complete.

It is left to you to complete the proof by verifying the uniqueness of the completion space $\widetilde{X}$.

3.10 Exercises

1. Let X be a metric space with metric ρ. Show that ρ_1, defined by the equation
$$\rho_1(x,y) = \min\{\rho(x,y), 1\},$$
is also a metric on X. If $X = \mathbb{R}$ with respect to the usual metric, then construct the open sphere relative to ρ_1.

2. Determine the relative topology that is induced by $\mathbb{R}$ (with respect to the usual topology) for the set of integers, $\mathbb{Z}$.

3. If (X, ρ) is a metric space, then prove that
$$\rho_1(x,y) = \rho(x,y) / [1 + \rho(x,y)],$$
is also a metric on X. Show that the two metric spaces (X, ρ) and (X, ρ_1) have the same open sets.

4. Let $X = \mathbb{R}^2$ and ρ be the usual metric on $\mathbb{R}^2$. Denote by O the point $(0,0)$. Define $\rho_1(x,y) = \rho(0,x) + \rho(0,y)$ for any $x = (x_1, x_2)$ and (y_1, y_2) in $\mathbb{R}^2$ and $x \neq y$, and $\rho_1(x,x) = 0\ \forall x \in \mathbb{R}^2$. Show that:

    (a) ρ_1 is a metric on $\mathbb{R}^2$;

    (b) All points other than O are open.

5. Let $(X_1, \rho_1), (X_2, \rho_2), \cdots (X_n, \rho_n)$ be metric spaces and let $X = X_1 \times X_2 \times \cdots \times X_n$. Define $\rho : X \times X \to [0, \infty)$ by $\rho(x,y) = \sqrt{\sum_{i=1}^{n} \rho_i(x_i, y_i)}$, where $x = (x_1, \cdots, x_n)$ and $y = (y_1, \cdots, y_n)$ in X. Show that ρ is a metric on X.

6. Let X be a metric space and let A, B be any two subsets of X. Show that $\overline{A \cap B} \subseteq \bar{A} \cap \bar{B}$. Give an example where the equality fails to hold.

7. Find a metric space which illustrates that $\overline{S_r(x)} \neq \{y \in X | \rho(x,y) \leq r\}$. (Hint: Let X be any set equipped with the discrete metric. Then $S_1(x) = \{y \in X | \rho(x,y) \leq r\} = \{x\}$. Hence $\overline{S_1(x)} = \{x\}$. On the other hand $\{y \in X | \rho(x,y) \leq 1\} = X$. Thus
$$\overline{S_r(x)} \neq \{y \in X | \rho(x,y) \leq r\}.)$$

8. Give a detailed proof that with respect to the usual metric $\mathbb{R}$ is a complete metric space.

9. Show that the set of integers, $\mathbb{Z}$, is a complete subspace of $\mathbb{R}$ (with respect to the usual metric). Identify the Cauchy sequences in $\mathbb{Z}$.

10. Give an example to show that completeness is not a topological property. (Hint: Consider $X = (-1, 1)$ and define a sequence $x_n = 1 - \frac{1}{n}$. Then X is not complete, however $(-1, 1)$ is homeomorphic to $\mathbb{R}$.)

11. Show that the set of rational numbers, $\mathbb{Q}$, is not homeomorphic to any complete metric space. Note that $\mathbb{Q}$ is neither an open nor a closed subset of $\mathbb{R}$.

12. Give an example to show that every subspace of a complete metric space is not necessarily complete.

13. Let (X, ρ) be a metric space and let $\{x_n\} \to x$ and $\{y_n\} \to y$. Show that $\lim_{n \to \infty} \rho(x_n, y_n) = \rho(x, y)$.

14. Prove that the following mappings from $\mathbb{R}$ into itself are contractive:

    (a) $f(x) = \sin x$;

    (b) $f(x) = |x|$.

15. Prove that the Cantor set is nowhere-dense in $[0, 1]$. Recall that the Cantor set is the set obtained by deleting a sequence of open sets, known as middle thirds, from $[0, 1]$. Precisely, $C = \bigcap_{n=1}^{\infty} E_n$, where $E_1 = [0, 1], E_2 = \left[0, \frac{1}{3}\right] \cup \left[\frac{2}{3}, 1\right], E_3 = \left[0, \frac{1}{9}\right] \cup \left[\frac{2}{9}, \frac{1}{3}\right] \cup \left[\frac{2}{3}, \frac{7}{9}\right] \cup \left[\frac{8}{9}, 1\right]$ and so on. Hence the Cantor set is of the first category.

16. Let $\{A_n\}$ be a sequence of open dense subsets of a metric space X. Show that $A = \bigcap_{n=1}^{\infty} A_n$ is also dense in X.

17. Using Baire's Category Theorem, prove that every countable complete metric space has an isolated point. (Recall that a point x of X is an isolated point iff x has a *nbd* that contains no other point of X, i.e., $\{x\}$ is open in x.)

18. Show that any subsequence of a Cauchy sequence is Cauchy.

19. Suppose that X is a metric space and that A is a countable subset of X. Show that A is of the first category in X iff A has no isolated point in X.

20. Show that the countable union of first category sets is of the first category.

21. Show that every subset of a first category set is first category.

22. Is it true that every open interval of $\mathbb{R}$ is of the second category? Justify your claim.

Chapter 4

Continuous Functions and Homeomorphisms

Some topologists regard Topology as the study of continuity in the most general context. As explained earlier, geometrically different figures are topologically the same if one can be deformed *continuously* into the other. Continuity of functions of real or complex variables is a familiar concept. For more general topological spaces a more precise definition is required. As always, the requirement remains that this definition be a generalization of continuity in the sense of functions of real variables. Therefore, we should be able to obtain the "$\epsilon - \delta$ definition" of Real Analysis as a special case of the general definition. That definition will be obtained later in this chapter.

A word about the term "homeomorphism". A "morphism" is a transformation, or mapping, from one space, say X_1, to another, say X_2. A "homeomorphism" is a structure preserving mapping, from the word "homo" meaning "same". When the mapping is from one topological space to another and the structure being preserved is continuity, it will be called a "homeomorphsim". If one geometrical figure is the homeomorphic image of the other the figures will be topologically equivalent. Already we have seen an example of topological equivalence in Theorem 3.4. There it was seen that squares and circles are topologically equivalent. That, and other examples dealt with so far, were in metric spaces. Here we shall deal with homeomorphisms in more general topological spaces.

4.1 Continuous Functions on Topological Spaces

To formulate the criterion for continuity let us start by considering an ordinary *dis*-continuous function like $f : \mathbb{R} \to \mathbb{R}$ given by

$$\left. \begin{aligned} f(x) &= x, & x < 1 \\ &= x + 1, & x \geq 1 \end{aligned} \right\}, \tag{4.1}$$

or $g : \mathbb{R} \to \mathbb{R}$ given by

$$\left. \begin{aligned} g(x) &= x, & x < 1 \\ &= x - 1, & x \geq 1 \end{aligned} \right\}. \tag{4.2}$$

These functions are displayed in Figure 4.1. In the former case there is a gap in the range while in the latter case the range gets overlapped. In either case there is a break in the graph. What is happening in both cases is that there are *nbds* of the image of $x = 1$ which have no corresponding *nbd* of $x = 1$. For continuity a *nbd* in one should correspond to a *nbd* in the other, so that "moving smoothly" in one space corresponds to "moving smoothly" in the other.

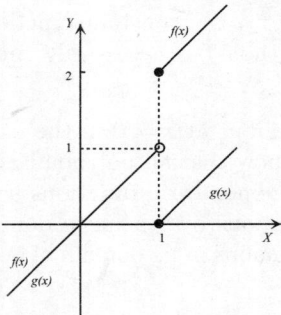

Figure 4.1: The functions $f(x)$ and $g(x)$ have broken graphs as is typical of discontinuous functions. Here points represented by filled circles are included while the point represented by the hollow circle is excluded from the graph. For $x < 1$ the two functions coincide while for $x \geq 1$ there are two distinct lines represented by the two functions.

In the general context, a function $f : (X_1, \tau_1) \to (X_2, \tau_2)$ is said to be *continuous at a point* $x_1 \in X_1$ if, for each *nbd* η_2 of $x_2 = f(x_1), \exists$ a *nbd* η_1 of x_1 s.t. $f(\eta_1) \subseteq \eta_2$. If f is continuous $\forall x_1 \in X_1$, we say that the function is *continuous*. Let us consider this definition applied to $f(x) = x^2$ at $x = 0$. Here the *nbd* of $x_1 = 0$ is $(-\epsilon, \epsilon) = \eta_1$ so that $f(\eta_1) = [0, \epsilon^2)$. Now $\eta_2 = (-\delta, \delta) \supseteq f(\eta_1)$ provided $\delta \geq \epsilon^2$. In other words, given η_2, we can always find $\epsilon \leq \sqrt{\delta}$ and hence find η_1 which satisfies the requirement that $f(\eta_1) \subseteq \eta_2$. (More generally, we could take $\eta_2 = (-\delta_1, \delta_2), (\delta_1, \delta_2 > 0)$.) Now consider $f(x)$ given by Eq. (4.1). Here $x_1 = 1$ and $x_2 = f(x_1) = 2$. Take $\eta_2 = (2-\delta, 2+\delta)$. Then for any $\eta_1 = (1-\epsilon, 1+\epsilon), f(\eta_1) = (1-\epsilon, 1) \cup (2, 2+\epsilon)$. Clearly, $f(\eta_1) \subsetneq \eta_2$ as $f(\eta_1) \smallsetminus \eta_2 = (1-\epsilon, 1)$ for $\epsilon \leq \delta \leq 1$. For $\delta < \epsilon$ there will be even more of $f(\eta_1)$ not contained in η_2. Similarly, for $g(x)$ given by Eq. (4.2), $x_1 = 1$ and $x_2 = 0$. Take $\eta_2 = (-\delta, \delta)$ and η_1 as before. Then $g(\eta_1) = (1-\epsilon, 1) \cup [0, \epsilon)$. Again $g(\eta_1) \subsetneq \eta_2$ as $f(\eta_1) \smallsetminus \eta_2 = (1-\epsilon, 1)$ for $\epsilon \leq \delta \leq \frac{1}{2}$ and $g(\eta_1) \smallsetminus \eta_2 \neq \phi$ for $\delta < 1$.

Another way of stating the requirement is that f is continuous at x_0 if, when $\eta_2 \in \mathfrak{N}(y_0)$ then $f^{-1}(\eta_2) \in \mathfrak{N}(x_0)$. Taking $\eta_2 \in \tau_2$ must clearly give $\eta_1 \in \tau_1$. It would be instructive to consider the inverse functions for Eq. (4.1) and Eq. (4.2) and verify that this formulation yields the same results as before. We leave this as an exercise for you.

Let us now proceed on to some examples of more unusual topological spaces, or more unusual situations in relatively common topological spaces.

Example 1. Any $f : (X_1, \tau_d) \to (X_2, \tau)$ is always continuous as every $U \subseteq X_1$ belongs to τ_d. (For simplicity consider $X_2 = X_1$ to be able to visualize the mapping more easily.) Hence every *nbd* of every point in (X_2, τ) has an open pre-image in (X_1, τ_d). Again any $f : (X_1, \tau) \to (X_2, \tau_i)$ is continuous as there is only one *nbd* of any point in X_2, namely X_2. Thus any *nbd* of the pre-image of that point in X_2 lies inside the pre-image of the *nbd* of that point, which is X_1.

Example 2. Any $f : (X, \tau_1) \to (X, \tau_2)$ s.t. $\tau_2 \subseteq \tau_1$ is continuous as there will then always exist smaller or equal *nbds* of the point than of its image. This is illustrated by taking $\tau_1 = \tau_u$ and $\tau_2 = \tau_c$ on $\mathbb{R}$. Now every element of τ_2 lies in τ_u but not *vice versa*, as $(0,1) \notin \tau_2$. Thus, for any point we can find a *nbd* in τ_u which is a proper subset of the *nbd* in τ_2. Clearly, the converse is not true, i.e., if $\tau_1 \subsetneq \tau_2$ then $f : (X, \tau_1) \to (X, \tau_2)$ must be discontinuous somewhere.

Example 3. If $f : (X_1, \tau_1) \to (X_2, \tau_2)$ is a constant function, i.e., $f(X_1) = \{a\} \in X_2 \ \forall x_1 \in X_1$, where a is a specific element of X_2, then f is necessarily continuous, as $f^{-1}(V) = X_1$ is an open *nbd* of all points of X_1.

We already saw that x^2 is continuous at $x = 0$ in the usual topology according to the above definition. It is a useful exercise to show that all polynomials are continuous for $(\mathbb{R}, \tau_u)$. Similarly, the exponential, hyperbolic and trigonometric functions $\sin x$ and $\cos x$ are continuous over $\mathbb{R}$ while functions like $\tan x$ are continuous over a finite domain, e.g., $(-\pi/2, \pi/2)$. The function $\sin^{-1} : [-1, 1] \to [-\pi/2, \pi/2]$ is continuous in its domain of definition. These, and similar functions, should be checked for continuity.

Theorem 4.1. $f : (X_1, \tau_1) \to (X_2, \tau_2)$ is continuous iff $\forall \ V \in \tau_2$, $f^{-1}(V) \in \tau_1$, i.e., the pre-image of every open set is an open set.

Proof. First assume that f is continuous and consider an arbitrary $x_1 \in U = f^{-1}(V)$, where V is open. By definition, V is a *nbd* of $x_2 = f(x_1)$. Since f is continuous, $\exists$ an open *nbd* G of x_1 s.t. $f(G) \subseteq V$. Thus $G \subseteq U$. Since this is true for arbitrary $x_1 \in U$, U must be open. Thus we have proved the forward implication. Now, take $U = f^{-1}(V)$ open. Thus $\exists$ an open *nbd* U of x_1 s.t. $f(U) \subseteq V$, which is open by definition and $x_2 = f(x_1) \in V$. Hence f is continuous at every arbitrary $x_1 \in U$. Thus f is continuous over all U. Since $\forall \ V \in \tau_2$ we have required $U = f^{-1}(V) \in \tau_1$, the function is continuous over the entire domain X_1. □

Consider a function mapping one topological space to the other when the second topology is given by a base. How can we check the continuity of the function? Since we have a relatively more manageable set to deal with, it should be easier to verify continuity here. This fact follows as a corollary of the above theorem.

Corollary 4.1.1. If $f : (X_1, \tau_1) \to (X_2, \tau_2)$ and τ_2 is generated by a topological bases β_2 then f is continuous iff $f^{-1}(B) \in \tau_1 \ \forall B \in \beta_2$.

Proof. For a bases $\beta_2 = \{\beta_\alpha | \alpha \in I\}$ of the topology τ_2, by definition any $V \subseteq X_2$ s.t. $V \in \tau_2$ can be written as $V = \bigcup_{\alpha \in J} B_\alpha, J \subseteq I$. Thus $f^{-1}(V) = \bigcup_{\alpha \in J} f^{-1}(B_\alpha)$, so that $f^{-1}(V)$ is open if each set $f^{-1}(B_\alpha)$ is. Hence the above theorem implies that f is continuous iff $f^{-1}(B_\alpha) \in \tau_1 \ \forall B_\alpha \in \beta_2$. □

This corollary can be used to show how functions acting on the same set with different topologies can be proved to be discontinuous. Consider $X_1 = X_2 = \mathbb{R}$, but use the usual topology as τ_1 and the discrete topology as τ_2. As we saw in an example given earlier, τ_d can be generated by the topological bases $\beta_2 = \{[a, b] | a, b \in \mathbb{R}\}$. Now consider even the identity map $f(x) = x$. Then while $[a, b] \in \tau_d$, $f^{-1}([a, b]) = [a, b] \notin \tau_u$, i.e., the pre-image of a τ_2 open set is not a τ_1 open set. Thus continuity of a function is intimately tied with the topology of the space. We also have the "dual" statement of the above corollary.

Corollary 4.1.2. $f : (X_1, \tau_1) \to (X_2, \tau_2)$ is continuous iff $f^{-1}(F)$ is closed $\forall F$ closed in τ_2.

Proof. By definition the complement of an open set is closed. Also

$$f^{-1}(V') = f^{-1}(X_2 \smallsetminus V) = f^{-1}(X_2) \smallsetminus f^{-1}(V) = X_1 \smallsetminus f^{-1}(V) = \left[f^{-1}(V)\right]'. \tag{4.3}$$

Consider any $F \subseteq X_2$ which is closed and define $G = f^{-1}(F) \subseteq X_1$. Now if F is closed F' is open. Thus, by the above theorem, f is continuous iff $f^{-1}(F')$ is open. By Eq. (4.3), then, $\left[f^{-1}(F)\right]'$ is open and hence $f^{-1}(F)$ is closed. Thus f is continuous iff $f^{-1}(F)$ is closed in X_1 $\forall F$ closed in X_2. $\square$

Before going on to discuss homeomorphisms and topological equivalence we will provide alternative characterizations of continuous functions and show, by some examples, how they are used.

4.2 Some Theorems About Continuous Functions

Instead of giving the requirements for continuity in terms of fully open or fully closed sets, it is useful to state them more generally in terms of arbitrary subsets, A_1, and their images $A_2 = f(A_1)$. This characterization of continuous functions is given by the following theorem.

Theorem 4.2. A function $f : (X_1, \tau_1) \to (X_2, \tau_2)$ is continuous iff

$$\forall A_1 \subseteq X_1, \ f(\bar{A}_1) \subseteq \overline{f(A_1)}. \tag{4.4}$$

Proof. We first prove necessity. For this purpose assume that f is continuous. Now consider $x \in \bar{A}_1$ and let V be an open *nbd* of $f(x)$. Then $f^{-1}(V) \subseteq X_1$. By Theorem 4.1, $f^{-1}(V) \subseteq X_1$ is open and contains x. Thus x is not a point on the boundary of $f^{-1}(V)$ and therefore lies in its interior. Hence, even if x lies only on the boundary of A_1, $f^{-1}(V) \cap A_1 \neq \phi$. Consequently $V \cap f(A_1) \neq \phi$. Consequently, $f(x) \in \overline{f(A_1)}$ for any arbitrary x. Now, for sufficiency assume that $f(\bar{A}_1) \subseteq \overline{f(A_1)}$ $\forall A_1 \subseteq X_1$ Let F be a closed set in X_2. Then, if $G = f^{-1}(F)$ is closed in X_1 we can use Corollary 4.1.2 of Theorem 4.1 to prove that f is continuous. If $\bar{G} = G$ then G is closed. Now $\bar{G} \supseteq G$. Thus if $\bar{G} \subseteq G$ then $\bar{G} = G$. Further $f(G) \subseteq F$. Hence $\overline{f(G)} \subseteq \bar{F} = F$ (as F is closed). Therefore if $x \in \bar{G}, f(x) \in f(\bar{G}) \subseteq \overline{f(G)}$ by our requirement. We thus have $f(x) \in F$ and so $f^{-1}(f(x)) = x \in f^{-1}(F) = G$. Hence $x \in \bar{G} \Rightarrow x \in G$ and so $\bar{G} \subseteq G$. We have, therefore, shown that if the condition is satisfied the inverse image of a closed set will be closed and so the mapping will be continuous. $\square$

A result that is often used in Analysis, for continuous functions is that the product of two functions, like the sum, is continuous. For the ratio of two functions it also holds except where the function in the denominator has a zero. This type of statement can not be generalized to spaces on which binary operations are not defined. However, we can compose functions acting on arbitrary topological spaces. The following theorem applies to this generalization.

Theorem 4.3. If $f : (X_1, \tau_1) \to (X_2, \tau_2)$ and $g : (X_2, \tau_2) \to (X_3, \tau_3)$, then the composite map $(g \circ f) : (X_1, \tau_1) \to (X_3, \tau_3)$ is continuous.

Proof. Consider $U \subseteq X_3$ s.t. $U \in \tau_3$. Since g is continuous, $g^{-1}(U) \in \tau_2$ and since f is continuous $f^{-1}\left[g^{-1}(U)\right] \in \tau_1$. By the basic definition $f^{-1}\left[g^{-1}(U)\right] = (g \circ f)^{-1}(U)$ and hence $(g \circ f)$ is continuous. □

We should, as mentioned at the start of this chapter, be able to derive the formal definition of continuity of a real valued function of a single real variable, from the general definition. This is given in the following theorem for metric spaces.

Theorem 4.4. $f : (X_1, \rho_1) \to (X_2, \rho_2)$ is continuous at $x_1 \in X_1$ iff for all real numbers $\epsilon > 0, \exists\, \delta > 0$ s.t. $\rho_1(x_1, y_1) < \delta \Rightarrow \rho_2(f(x_1), f(y_1)) < \epsilon$.

Proof. Recall that a metric induces a topology whose base is the collection of "open spheres" defined by the metric. If f is continuous, construct for a given $\epsilon > 0$, the ρ_2 open sphere, $S_\epsilon^2(f(x_1))$ of X_2. By the continuity of f, $f^{-1}\left[S_\epsilon^2(f(x_1))\right] \ni x_1$ and is open in X_1. Thus $\exists\, \epsilon > 0$ s.t. $S_\delta^1(x_1) \subseteq f^{-1}\left[S_\epsilon^2(f(x_1))\right]$. Now consider $y_1 \in S_\delta^1(x_1)$, Hence $\rho_1(x_1, y_1) < \delta$ and $\rho_2(x_2, y_2) = \rho_2(f(x_1), f(y_1)) < \epsilon$ as $y_2 \in S_\epsilon^2(x_2)$. Conversely, if the condition holds, consider $V \ni x_2$, Then $\exists\, S_\epsilon^2(x_2) \subseteq V$ which has radius ϵ and is centered at x_2. We have assumed that $\exists\, \epsilon > 0$ s.t. $S_\delta^1(x_1)$ centered at x_1 satisfies the requirement that $f\left(S_\delta^1(x_1)\right) \subseteq S_\delta^{-2}(f(x_1))$, which is open. Therefore $f^{-1}(V)$ is an open *nbd* of x_1 and hence is continuous. □

Consider the simplest example for Theorem 4.4, that $X_1 = X_2 = \mathbb{R}$ and $\rho_1 = \rho_2 = \rho_u$. In that case we get the usual "$\epsilon - \delta$" definition of continuity of functions of 1 real variable. Equally, taking $X_1 = \mathbb{R}^n, X_2 = \mathbb{R}, \rho_1 = \rho_2 = \rho_u$, we get the extension to functions of several variables. Still more generally, taking $X_1 = \mathbb{R}^n, X_2 = \mathbb{R}^m, \rho_1 = \rho_2 = \rho_u$ we get the extension to vector valued functions of several variables which are real. It is obvious that the definition can, in fact, be generalized to *any* metric space by Theorem 4.4. Thus, for example, complex vector valued functions of several complex variables can be tested for continuity in the same way. In that case, of course, we would also have to worry about the complex structure being preserved. We will not bother, here, with the ramifications of such functions. Instead, we turn to a characterization of continuity in terms of convergent sequences that is also used in Real Analysis.

Theorem 4.5. $f : (X_1, \rho_1) \to (X_2, \rho_2)$ is continuous iff for all convergent sequences $x_n \to \overline{x}$ in X_1, the sequence $f(x_n) \to f(\overline{x})$ in X_2 is convergent.

Proof. Consider $S_\epsilon^2(f(\overline{x}))$ for a continuous f. Then $\exists\, \delta > 0$ s.t. $f^{-1}(S_\epsilon^2(f(\overline{x})) \subseteq S_\delta(\overline{x})$ (as in the previous theorem). Since $x_n \to \overline{x}, \exists N \in \mathbb{N}$ s.t. $x_n \in S_\delta(\overline{x})$, $\forall n \geq N$. Then $f(x_n) \in S_\epsilon^2(f(\overline{x}))$, $\forall n \geq N$, i.e., $f(x_n)$ converges to $f(\overline{x})$. To prove sufficiency, assume that the sequence $f(x_n)$ converges to $f(\overline{x})$ as x_n converges to $\overline{x}$. Suppose that f is not continuous at some point $\overline{x} \in X_1$. Then $\exists\, V \in \tau_2$ and $V \ni f(\overline{x})$ such that in every open sphere $S_{1/n}^1(\overline{x})$ there exists at least one point x_n^* (say) whose image, $f(x_n^*) \notin V$. Therefore we can extract the sequence $\{x_n^*\}$ from the sequence of open spheres such that $x_n^* \in S_{1/n}^1(x)$ and $f(x_n^*) \notin V$. Thus $x_n^* \to \overline{x}$ but $f(x_n^*) \not\to f(\overline{x})$, in contradiction to our assumption that the sequence converges. By *reductio ad absurdum* we conclude that f must be continuous. □

Let us look at some examples of the application of the above theorems.

Example 1. Consider $X_1 = X_2 = \mathbb{R}$, and $\tau_1 = \tau_2 = \tau_u$ and the function

$$\begin{aligned} f(x) &= 1 \quad \text{if} \quad x \in \mathbb{Q} \\ &= 0 \quad \text{if} \quad x \in \mathbb{Q}' \end{aligned} \Bigg\}. \tag{4.5}$$

Remember that $\overline{\mathbb{Q}} = \overline{\mathbb{Q}'} = \mathbb{R}$. Consider any $x \in \mathbb{Q}$ and assume that f is continuous there. Now $x \in \overline{\mathbb{Q}'}$ and so, by Theorem 4.2, $f(x) \in f(\overline{\mathbb{Q}'}) \subseteq \overline{f(\mathbb{Q}')}$. Now we have $f(\mathbb{Q}') = \{0\}$ and $\overline{f(\mathbb{Q}')} = \overline{\{0\}} = \{0\}$, as a singleton set is closed. But $f(x) = 1$. Hence we have a contradiction. Thus the assumption that f is continuous at x is false $\forall x \in \mathbb{Q}$. Hence $f(x)$ is never continuous!

Example 2. Consider $X_1 = X_2 = \mathbb{R}$, but $\tau_1 = \tau_c$ and $\tau_2 = \tau_u$, with the function defined by $f(x) = x$, $\forall x \in \mathbb{R}$. Again assume that f is continuous and use Theorem 4.2. Then $f(\overline{\mathbb{Z}}) \subseteq \overline{f(\mathbb{Z})}$. In τ_c, $\overline{\mathbb{Z}} = \mathbb{R}$ while in τ_u, $\overline{\mathbb{Z}} = \mathbb{Z}$. Thus if f is continuous $\mathbb{R} \subseteq \mathbb{Z}$. Since this is impossible f is not continuous, even in the simple case of an identity function, because the topologies are different.

Example 3. For any topological space consider $A \subseteq X$ and define the characteristic function $\chi_A : X \to \mathbb{R}$ by

$$\begin{aligned} \chi_A &= 1 \quad \text{if} \quad x \in A \\ &= 0 \quad \text{if} \quad x \in A' \end{aligned} \Bigg\}. \tag{4.6}$$

We require χ_A to be continuous and ask for the topology. Now $\chi_A(X) = \{0, 1\}$ and hence it has the discrete topology. Since $\{1\}$ is open, by Theorem 4.1, $\chi_A^{-1}(\{1\}) = A$ is open. Also, since $\{1\}$ is closed, $\chi_A^{-1}(\{1\}) = A$ is closed (by Corollary 4.1.2). Hence, for χ_A to be continuous A must be both open and closed in the topology on X.

Example 4. $f : \mathbb{R} \to \mathbb{R}$ given by $f(x) = 2x + 1$ with the usual topology is obviously continuous. This may be seen by the use of Theorem 4.5. Consider a sequence $\{x_n\}$ which converges to $\overline{x} \in \mathbb{R}$. Thus, for every $\epsilon > 0$ $\exists N$ s.t. $|x_n - \overline{x}| < \epsilon$, $\forall n \geq N$. Now $|f(x_n) - f(\overline{x})| = |(2x_n + 1) - (2\overline{x} + 1)| = 2|x_n - \overline{x}|$. Thus $|f(x_n) - f(\overline{x})| < 2\epsilon$. Hence taking $\delta = \epsilon/2$ we have found that $f(x_n)$ converges to $f(\overline{x})$.

Example 5. For the same topological spaces take $f(x) = |x|$. Now $||x| - |y||$ $\leq |x - y|$ as is obvious by considering x and y having the same and opposite signs. Again use Theorem 4.5. If $x_n \to \overline{x}$, $|x_n - \overline{x}| \to 0$ as $n \to \infty$. Thus $|f(x_n) - f(\overline{x})| = ||x_n| - |\overline{x}|| \leq |x_n - \overline{x}|$. Hence $|f(x_n) - f(\overline{x})| \to 0$ as $n \to \infty$. This is shown in Figure 4.2.

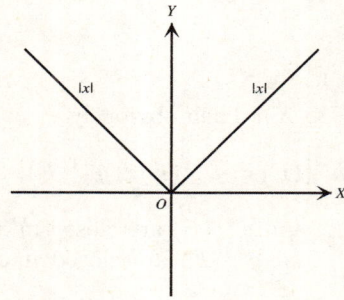

Figure 4.2: The function $f(x) = |x|$ is depicted for some domain. It is clearly continuous everywhere, including $x = 0$.

Example 6. The *projection mapping* $p_1 : \mathbb{R}^2 \to \mathbb{R}$ given by $p_1(x_1, x_2) = x_1$, essentially takes the "shadow" of a vector in the plane on to the first axis. (More generally the projection could be on to any line, e.g., along X_2 or along $x_2 = x_1$.) Now $\rho_1(\mathbf{x}, \mathbf{y})$ is taken to be the "diamond metric" $|x_1 - y_1| + |x_2 - y_2|$. Let $\{\mathbf{x}^{(n)}\}$ be a sequence in $\mathbb{R}^2$ which converges to $\overline{\mathbf{x}}$. Since $\forall \epsilon > 0, \exists N \ s.t. \ \rho\left(\mathbf{x}^{(n)}, \overline{\mathbf{x}}\right) = \left|x_1^{(n)} - \overline{x}_1\right| + \left|x_2^{(n)} - \overline{x}_2\right| < \epsilon, \ \forall n \geq N, \ \left|p_1(\mathbf{x}^{(n)}) - p_1(\overline{\mathbf{x}})\right| = \left|x_1^{(n)} - \overline{x}_1\right| < \epsilon, \ \forall n \geq N$. Hence the projection mapping is continuous. One can have a projection mapping $p_m : \mathbb{R}^n \to \mathbb{R}^m$, $m < n$ given by $p_m(x_1, \cdots, x_m, x_{m+1}, \cdots, x_n) = (x_1, \cdots, x_m)$ and repeat the above argument to show that it is continuous. Notice that $p_m^2 = p_m$ in that we could have regarded $\mathbb{R}^m$ as a subset of $\mathbb{R}^n$ and applied p to it again. It is called an *idempotent function* because of this property.

Example 7. The *translation mapping* $f : \mathbb{R}^n \to \mathbb{R}^n$ given by $f(\mathbf{x}) = \mathbf{x} + \mathbf{a} = \{x_i + a_i\}$ ($i = 1, \cdots, n$), is continuous as may be seen by constructing a sequence $\{\mathbf{x}^{(n)}\}$ which converges to $\overline{\mathbf{x}}$. The details of the proof are left as an exercise for you.

Example 8. Define $F_t : (C([a,b]), \rho_1) \to (\mathbb{R}, \rho_1)$ at a given $t \in [a,b]$ by $F_t(f) = f(t)$, where ρ_1 is the continuous generalization of the square metric. Construct the sequence $\{f_n\}$ in $C([a,b])$ s.t. $f_n \to f$, i.e., $\forall \epsilon > 0 \ \exists N(\epsilon)$ s.t. $\rho(f_n, f) = \max_{a \leq t \leq b} |f_n(t) - f(t)| < \epsilon \ \forall n \geq N(\epsilon)$. Then we have $|F_t(f_n) - F_t(f)| = |f_n(t) - f(t)| \leq \max_{a \leq t \leq b} |f_n(t) - f(t)| < \epsilon, \forall n \geq N$. Hence $F_t(f_n) \to F_t(f)$ and so F_t is continuous.

Example 9. The "dot product" on $\mathbb{R}^n$ is defined by $f : \mathbb{R}^n \to \mathbb{R}$ given by $f(\mathbf{x}) = \mathbf{a} \cdot \mathbf{x} = \sum_{i=1}^n a_i x_i$. This is a continuous function. Once again the proof is left to you as a simple exercise.

A very important technique for dealing with functions on curved spaces is to *patch* them together. That this can be done seems obvious. However, it needs to be proved that it can, and we need to spell out the conditions which must be met for this to be possible. This requirement is met by the following "*pasting lemma*".

Theorem 4.6. If $X_1 = A \cup B$, where A and B are closed sets and $f : A \to X_2$, $g : B \to X_2$ are both continuous with $f(x) = g(x), \ \forall x \in A \cap B$ then $h : X_1 \to X_2$ defined by

$$h(x) \begin{array}{l} = f(x) \ \text{if} \ x \in A \\ = g(x) \ \text{if} \ x \in B \end{array} \Bigg\}. \tag{4.7}$$

is a continuous function.

Proof. Consider a closed subset $F \subseteq X_2$. Then obviously

$$h^{-1}(F) = f^{-1}(F) \cup g^{-1}(F), \tag{4.8}$$

since f and g are continuous $f^{-1}(F)$ and $g^{-1}(F)$ are closed in A and B respectively. Also, since A and B are closed in X_1, $f^{-1}(F)$ and $g^{-1}(F)$ are closed in X_1. Thus, by Corollary 4.1.2 or Theorem 4.1, h is continuous. □

It should be clear that we could have required A and B to be both open and used Theorem 4.1 instead of its corollary. However, we can not have them neither open nor closed, or one of each.

For example $f(x)$ or $g(x)$ of Figure 4.1 are *not* continuous, though they are both continuous for $x \in (-\infty, 1)$ and $x \in [1, \infty)$. For continuity to be meaningful h must be defined on all of X_1. For both open it is necessary that $A \cap B \neq \phi$. (Incidentally the procedure of matching f and g over $A \cap B$ is called *patching* in geometry instead of *pasting*.)

Example 10. Consider $h : \mathbb{R} \to \mathbb{R}$ given by

$$\left. \begin{array}{ll} h(x) &= 2x \quad \text{for} \quad x \leq 0 \\ &= x^2 \quad \text{for} \quad x \geq 0 \end{array} \right\}, \tag{4.9}$$

has $A = (-\infty, 0], B = [0, \infty)$ so that $A \cap B = \{0\}$, $f(0) = 0$, $g(0) = 0$. Hence the conditions are met and $h(x)$ is continuous, see Figure 4.3(a). Instead, take

$$g(x) = x^2 + 1 \quad \text{for} \quad x \geq 0, \tag{4.10}$$

so that $f(0) = 0$, $g(0) = 1$. Here the conditions are *not* met and the function is discontinuous, as is apparent from Figure 4.3(b).

(a)   (b)

Figure 4.3: (a) $h(x)$ defined as $f(x)$ for $x \leq 0$ and $g(x)$ for $x \geq 0$ can be pasted together at $x = 0$ to get a continuous function. (b) $f(x)$ and $g(x)$ can not be pasted together at $x = 0$ as they do not agree there.

Example 11. The *step function* is defined by

$$\left. \begin{array}{ll} h(x) &= -1 \quad \text{for} \quad x \leq 0 \\ &= +1 \quad \text{for} \quad x > 0 \end{array} \right\}. \tag{4.11}$$

Clearly, like $g(x)$ in the previous example, $h(x)$ is discontinuous at $x = 0$ (see Figure 4.4). Quite often $h(x)$ is left undefined at $x = 0$ and written as

$$\left. \begin{array}{ll} h(x) &= -1 \quad \text{for} \quad x < 0 \\ &= +1 \quad \text{for} \quad x > 0 \end{array} \right\}, \tag{4.12}$$

so that $h(0)$ is left out. The function remains discontinuous since it is not defined at $x = 0$.

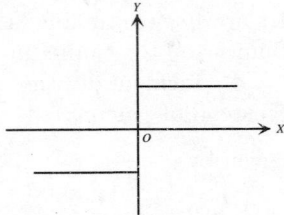

Figure 4.4: The step function is discontinuous at $x = 0$.

4.3 Homeomorphisms

We had previously stated that homomorphisms are structure preserving mappings. Though correct, the definition is incomplete, as it does not state what is entailed in "structure preserving". A one-to-one mapping from one set *into* another is called an *injective* mapping. If the mapping is *onto* the other set, it is called *surjective*. These mappings are depicted in Figure 4.5. If a mapping is both into and onto it is called *bijective*. If there is an associated algebraic structure, likea binary operation, which is carried through by the mapping, it is called an *isomorphism* (from the Greek "iso" for "same" and "morph" for "shape"). If it is also invertible, i.e., if f^{-1} exists, it is called a *homomorphism*. If we further have continuity of the mapping it is called a *homeomorphism*. The structure preserved here is the topology. Thus open sets in one *space* are mapped into open sets in the other space and *vice versa*. If (X_1, τ_1) can be mapped to (X_2, τ_2) by a homeomorphism they are said to be *homeomorphic* to each other, denoted by $(X_1, \tau_1) \cong (X_2, \tau_2)$. Often the topology is implicitly given by the statement $X_1 \cong X_2$.

(a)   (b)

Figure 4.5: (a) The mapping $f : A \to B$ is said to be *injective* or *into*. Not every element of B is necessarily an image of an element in A. For instance there does not exist a pre-image of y in A, i.e., $f^{-1}(y) \not\ni A$. (b) The mapping $g : A \to B$ is said to be *surjective* or *onto*. Not necessarily every element of A has an image in B. For instance there is no image of x in B, i.e., $f(x) \not\ni B$. However, $f : A \to D$ and $g : C \to B$ are both into and onto. Here, every element in the former set has an image in the latter set and every element in the latter set is an image of an element in the former set.

Example 1. The identity function $f : X \to X$ given by $f(x) = x \; \forall x \in X$ is a homeomorphism. Also, consider $X_1 = (0, \infty)$, $X_2 = (-\infty, 0)$ and define $f : X_1 \to X_2$ given by $f(x) = -x$. Then $f(x)$ is a homeomorphism. Similarly $f : (-\pi/2, \pi/2) \to \mathbb{R}$ given by $f(x) = \tan x$ is a homeomorphism with $f^{-1}(x) = \tan^{-1} x$. Similarly $f : \mathbb{R} \to (0, \infty)$ given by $f(x) = e^x$ is a homeomorphism with $f^{-1} : (0, \infty) \to \mathbb{R}$ given by $f^{-1}(x) = \ln x$. Again $f : \mathbb{R} \to \mathbb{R}$ given by $f(x) = 2x + 1$ has $f^{-1} : \mathbb{R} \to \mathbb{R}$

given by $f^{-1}(y) = \frac{1}{2}(y-1)$ so that $f\left(f^{-1}(y)\right) = f\left(\frac{1}{2}(y-1)\right) = 2 \times \frac{1}{2}(y-1) + 1 = y$. All are obviously continuous.

Example 2. The *translation* $f : \mathbb{R}^n \to \mathbb{R}^n$ given by $f(\mathbf{x}) = \mathbf{x} + \mathbf{a}$, is a homeomorphism as it is invertible. $f^{-1} : \mathbb{R}^n \to \mathbb{R}^n$ is given by $f^{-1}(\mathbf{y}) = \mathbf{y} - \mathbf{a}$ so that $f\left(f^{-1}(\mathbf{y})\right) = f(\mathbf{y}-\mathbf{a}) = \mathbf{y}-\mathbf{a}+\mathbf{a} = \mathbf{y}$. Now consider a transformation $f : \mathbb{R}^n \to \mathbb{R}^n$ given by $\widehat{x}_i = f_i(x_j) = \sum_{j=1}^n A_{ij} x_j$. We write the column vector as $\mathbf{x}^t$. Then the *length* of the vector, $\mathbf{x}^t \cdot \mathbf{x}$, will remain invariant if the transformation is a rotation. Thus, for a *rotation*

$$\sum_{i=1}^n \widehat{x}_i \widehat{x}_i = \sum_{i,j,k=1}^n x_i A_{ij}^t A_{jk} x_k = \sum_{k=1}^n x_k x_k. \tag{4.13}$$

Hence, we require that

$$\sum_{i=1}^n A_{ij}^t A_{jk} = \delta_{ik}, \tag{4.14}$$

where δ_{ik} is the *Kronecker delta*, being 1 along the diagonal and 0 off-diagonal, i.e.,

$$\left. \begin{array}{ll} \delta_{ik} &= 1 \text{ if } k = i \\ &= 0 \text{ if } k \neq i \end{array} \right\}. \tag{4.15}$$

In other words δ_{ik} is the identity matrix in index notation. Such transformations are called *orthogonal transformations*. Since the product of determinants of a matrix is the determinant of the product, from Eq. (4.14) $\left[\det\left(A_{ij}\right)\right]^2 = 1$, for orthogonal matrices. Henceforth we shall write $\det(A)$ as $|A|$. For the matrix to represent a rotation, the determinant must be $+1$ and the -1 determinant only corresponds to a reflection. These may be easily understood in the 2-dimensional example, where the rotation matrix is

$$A = \begin{pmatrix} \cos\theta & \sin\theta \\ -\sin\theta & \cos\theta \end{pmatrix}, \tag{4.16}$$

and, for example, reflections in the X-axis and Y-axis are given by

$$R_1 = \begin{pmatrix} 1 & 0 \\ 0 & -1 \end{pmatrix}, R_2 = \begin{pmatrix} -1 & 0 \\ 0 & 1 \end{pmatrix}. \tag{4.17}$$

Notice that $|A| = 1$ while $|R_1| = |R_2| = -1$. Both translations and rotations are continuous. Hence they are homeomorphisms. However, since the reflections are discontinuous they are not homeomorphisms.

Example 3. $f : \mathbb{R} \to \mathbb{R}$ given by $f(x) = x^2$ is *not* a homeomorphism as f is not bijective. Again $f : [0, 2\pi] \to [-1, 1]$ given by $f(x) = \sin x$ is *not* invertible. Thus, even though these functions are continuous they are *not* invertible and hence are not homeomorphisms. (To see why consider the graph of $\sin x$ in the given domain, shown in Figure 4.6(a), and reflect it in the line $y = x$ (i.e., interchange x and y) to invert it as shown in Figure 4.6(b). Since every vertical line in the open domain $(-1, 1)$ cuts the curve more than once, there is no unique inverse for $f(x)$ and hence it is not invertible. Similarly for the parabola given above. [Make the graph and reflect.])

Figure 4.6: (a) Graph of $\sin x$. (b) Graph of $\sin^{-1} x$.

However, $f : [0, \infty) \to [0, \infty)$ given by $f(x) = x^2$ *is* a homeomorphism and $f(x) = \sin x$ with domain $[-\pi/2, \pi/2]$ and range $[-1, 1]$ *is* a homeomorphism, as they are both continuous.

Example 4. $f : [0, 1) \to [1, \infty)$ given by $f(x) = (1-x)^{-1}$ is invertible. To find it put $f(x) = y$ and solve to find x as a function of y. Thus $1 - x = 1/y$ or $x = (y-1)/y$. Thus $f^{-1}(y) = (y-1)/y$ and $f(f^{-1}(y)) = (1 - (y-1)/y)^{-1} = (1 - 1 + 1/y)^{-1} = (1/y)^{-1} = y$. Prove for yourself whether it is continuous in the domain or not.

Example 5. Consider $f : (\mathbb{R}, \tau_d) \to (\mathbb{R}, \tau_u)$ given by $f(x) = x$. Clearly f is invertible and $f^{-1}((a,b)) = \bigcup_{x \in (a,b)} \{x\}$ is open in the discrete topology. Thus f is obviously continuous. However $g = f^{-1} : (\mathbb{R}, \tau_u) \to (\mathbb{R}, \tau_d)$ is *not* continuous since $\{x\} \in \tau_d$ but $g^{-1}(\{x\}) = \{x\} \notin \tau_u$ as $\{x\}$ is closed in $\mathbb{R}$ with the usual topology. Hence f is *not* a homeomorphism.

Example 6. As shown earlier using the stereographic projection $\mathbb{S}^1 \smallsetminus \{(0, 1)\} \cong \mathbb{R}$. Try proving this by constructing the algebraic representation of the homeomorphism given by the stereographic projection. Similarly, for $\mathbb{S}^2 \smallsetminus \{(0, 0, 1)\} \cong \mathbb{R}^2$. Here S^1 is the circle $\{(x,y) | x^2 + y^2 = 1\}$ and S^2 is the surface of the sphere $\{(x, y, z) | x^2 + y^2 + z^2 = 1\}$.

Example 7. Consider the mapping $f : (X, \tau_u) \to (\mathbb{R}, \tau_d)$. Regardless of how it is constructed, if $X = A \cup B$ s.t. $A, B \in \tau_u, A \cap B = \phi$ and $f(A), f(B) \subseteq \mathbb{R}$, f can not be a homeomorphism, i.e., for such an X, $X \not\cong \mathbb{R}$. This is so because $A \cap B = \phi$ and $B \in \tau_u \Rightarrow \bar{A} \cap B = \phi$. Thus a point on ∂A can not lie inside B. Take $x \in \partial A$ and construct a *nbd* $\eta(f(x))$. Clearly, $f^{-1}(\eta(f(x))$ does not exist as it would need to include x. This may be seen more concretely by considering $X = (0,1) \cup (2,3)$. For any mapping $f : X \to \mathbb{R}$ we could look for the image of 1 in $\mathbb{R}$ and not be able to find the inverse image in X. In fact we could even take $X = (0, 1) \cup (1, 2)$. Sets A and B satisfying the above property give, as their union, a *disconnected space* X, while $\mathbb{R}$ is a *connected space*. This point is discussed in more detail in Chapter 6.

4.4 Open and Closed Continuous Functions

A priori there is no reason why the *image* of an open set under a continuous mapping should be open. All we are guaranteed is that the *pre-image* of an open set will be open. A continuous function which maps open sets to open sets is said to be *open*. Thus $f : (X_1, \tau_1) \to (X_2, \tau_2)$ s.t. $\forall V_1 \in \tau_2$, $f^{-1}(V_1) = U_1 \in \tau_1$ and $\forall U_2 \in \tau_1$, $f(U_2) = V_2 \in \tau_2$ is a *continuous open function* or an *open continuous function*. The same remarks apply to closed sets being mapped into closed sets. Functions that map closed sets into closed sets and are continuous, are called *closed*.

Example 1. The projection mapping $p_1 : (\mathbb{R}^2, \tau_u) \to (\mathbb{R}, \tau_u)$ given by $p_1(x_1, x_2) = x_1$ is an open mapping since the projection of any open sphere $S_r(x_1, x_2)$ is $(x_1 - r_1, x_1 + r)$, see Figure 4.7. This mapping is *not* closed. The reason is not because closed sets are not mapped to closed sets but because the image of infinite unions of closed sets may not be closed. It would be a useful exercise for you to show that a hyperbola is an example of a closed subset of $\mathbb{R}^2$ whose projection is not closed, but open.

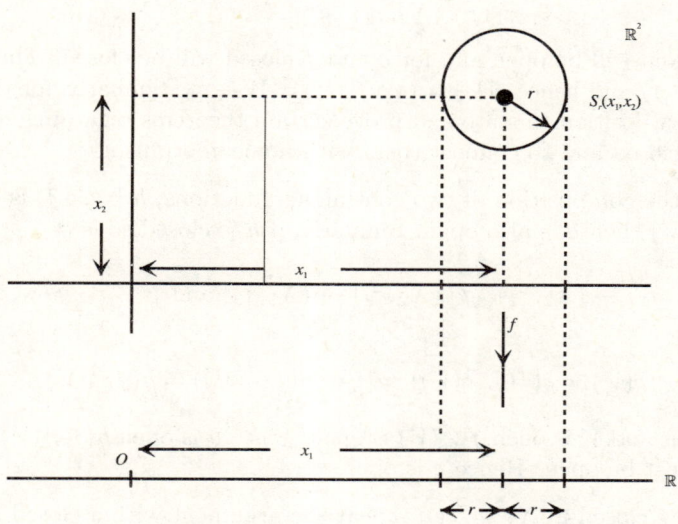

Figure 4.7: The function p_1 projects $\mathbb{R}^2$ to $\mathbb{R}$. Then open (closed) sphere $S_r(x_1, x_2)(\overline{S}_r(x_1, x_2))$ is mapped to the open (closed) interval $(x_1 - r, x_1 + r)([x_1 - r, x_1 + r])$. Thus the mapping is open. However, since the infinite union of closed sets is not necessarily mapped into closed sets, it is not a closed mapping.

Example 2. $f : (\mathbb{R}, \tau_u) \to (\mathbb{R}, \tau_u)$ given by

$$f(x) = (1 + x^2)^{-1} \quad \forall x \in \mathbb{R}, \tag{4.18}$$

is not open or closed as the image of $\mathbb{R}$ is $(0, 1]$, which is clearly neither open nor closed.

Example 3. $f : (X_1, \rho_1) \to (X_2, \rho_2)$ given by $f(x) = y_0 \;\forall x \in X_1$ and where $y_0 \in X_2$ is unique, is the constant function. This is clearly not an open mapping as $f(S_r(x)) = \{y_0\}$. Thus an open set has a closed image. It is a closed mapping as the image of every closed set is the closed set $\{y_0\}$.

Example 4. $f : (\mathbb{R}, \tau_u) \to (\mathbb{R}, \tau_c)$ is continuous but neither open nor closed for $f(x) = x$. This is because, for example $f((0, 1)) = (0, 1)$, but $(0, 1) \in \tau_u$ while $(0, 1) \notin \tau_c$. Similarly, $f([0, 1]) = [0, 1]$ and while $[0, 1]$ is closed in the usual topology it is not closed in the co-finite topology.

Example 5. $f : (\mathbb{R}, \tau_u) \to ((-\pi/2, \pi/2), \tau_u)$ given by $f(x) = \tan^{-1} x$ is both open and closed. Also, conversely $g : ((-\pi/2, \pi/2), \tau_u) \to (\mathbb{R}, \tau_u)$ given by $f(x) = \tan x$ is both open and closed.

All functions which map the whole of $\mathbb{R}$ onto the whole of $\mathbb{R}$ (with the usual topology) and always increase or always decrease are both open and closed. For example $\sinh x$, $x^3 + 3x$, etc. Again $f(x) = \tanh^{-1} x$ or $\tanh x$ are open and closed in the appropriate domains and ranges, i.e., $\tanh : (\mathbb{R}, \tau_u) \to ((-1,1), \tau_u)$ and $\tanh^{-1} : ((-1,1), \tau_u) \to (\mathbb{R}, \tau_u)$.

It is clear that the composition of open mappings is open and of closed mappings is closed, i.e., if
$$f : (X_1, \tau_1) \to (X_2, \tau_2), \ g : (X_2, \tau_2) \to (X_3, \tau_3), \tag{4.19}$$
$g \circ f$ for g and f open will be open and for g and f closed will be closed. This is obvious since if $U \in \tau_1, f(U) = V \in \tau_2$ and hence $g(V) = (g \circ f)(U) = W \in \tau_3$. Similarly, for the closed subsets of X_1. We are now in a position to state and prove certain theorems regarding continuous functions which are open or closed and to connect them with homeomorphisms.

Theorem 4.7. If the composition of two continuous functions, $h = g \circ f$, is open and f is *onto* (i.e., $f(f^{-1}(V)) = V$) then g is also open. Similarly, if h is closed so is g.

Proof. Let $f : (X_1, \tau_1) \to (X_2, \tau_2)$, $g : (X_2, \tau_2) \to (X_3, \tau_3)$ and $V \in \tau_2$. Now, since f is onto (also called *surjective*),
$$g(V) = g(f(f^{-1}(V))) = (g \circ f)(f^{-1}(V)) = h(f^{-1}(V)), \tag{4.20}$$
since f is continuous and V is open $f^{-1}(V)$ is open. Since h is open $h(f^{-1}(V))$ is open. Thus, for an open V, $g(V)$ must be open. Hence g is open. $\square$

Instead, taking f closed in X_2 we can repeat the argument with a closed h giving a closed g. This is shown diagramatically in Figure 4.8.

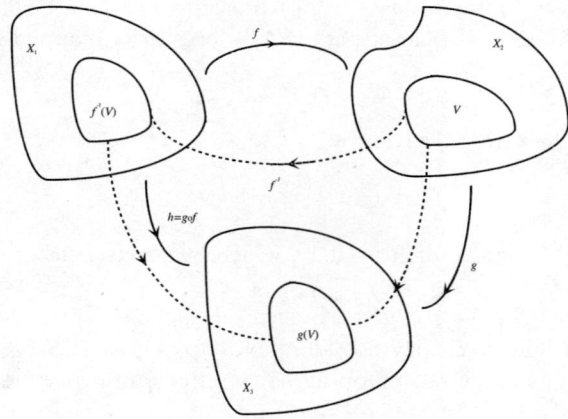

Figure 4.8: Since $f : X_1 \to X_2$ is invertible f^{-1} can map $V \subseteq X_2$ to $f^{-1}(V) \subseteq X_1$ and g maps V to $g(V) \subseteq X_3$. Then $h = g \circ f : X_1 \to X_3$ acts on $f^{-1}(V)$ to give $g(V)$. All sets: V, $f^{-1}(V)$, $g(V)$; are open for g and h open or closed for f and h closed.

Theorem 4.8. If the composition of two continuous functions, $h = g \circ f$, is open and g is one-to-one and *into* (i.e., $g^{-1}(g(V) = V)$ then f is also open. Similarly, if h is closed so is f.

Proof. Consider, here, $U \in \tau_1$. By the requirement on g (*injectivity*),
$$g^{-1}(h(U)) = g^{-1}(g(f(U))) = f(U). \tag{4.21}$$
Since h is open and U is open $h(U) \in \tau_3$. Now as g is continuous $g^{-1}(h(U)) \in \tau_2$. Hence $f(U)$ is open in X_2 for any open U in X_1. Hence f is open. □

Again, the corresponding argument with F closed in X_1 applies for closed h giving closed f. This is shown diagramatically in Figure 4.9.

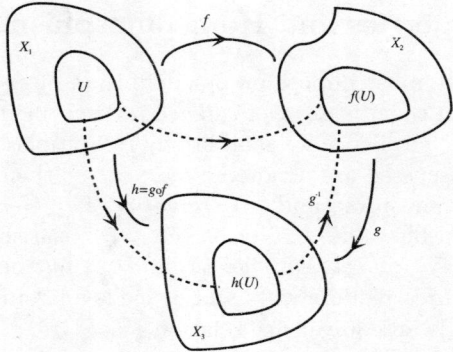

Figure 4.9: As g is one-to-one, into and invertible g^{-1} can map the image of $U \subseteq X_1$ under $h, h(U) \subseteq X_3$ to $g^{-1}(h(U)) = f(U) \subseteq X_2$. Again all sets: U, $h(U)$, $f(U)$ are open for h open or closed for h closed.

Theorem 4.9. f is closed iff it commutes with the closure operation, i.e., $f(\bar{A}_1) = \overline{f(A_1)}$ $\forall A_1 \subseteq X_1$.

Proof. First assume f is closed. By Theorem 4.2, $f(\bar{A}_1) \subseteq \overline{f(A_1)}$. Also $f(A_1) \subseteq f(\bar{A}_1)$ and $\overline{f(A_1)} \subseteq \overline{f(\bar{A}_1)}$. Now $f(\bar{A}_1)$ is a closed set since f is closed. Thus $f(\bar{A}_1) = \overline{f(\bar{A}_1)}$. Therefore $f(\bar{A}_1) \subseteq \overline{f(A_1)} \subseteq f(\bar{A}_1)$ or $\underline{f(\bar{A}_1)} = \overline{f(A_1)}$. Now assume the condition holds. Then $f(\overline{A_1}) = \overline{f(A_1)}$ and $f(\overline{A_1}) = f(A_1)$. Thus $f(A_1) = \overline{f(A_1)}$ or f is closed. □

We are finally in a position to connect the requirements of openness and closedness with homeomorphisms. This is achieved by the following theorem.

Theorem 4.10. A homeomorphism is both open and closed.

Proof. Write the inverse of the homeomorphism f as g. Thus $\forall A_1 \subseteq X_1, f(A_1) = g^{-1}(A_1)$. If A_1 is open (or closed) in X_1 then $g^{-1}(A_1)$ is correspondingly open (or closed) in X_2 as f is a homeomorphism. Hence $f(A_1)$ is open (or closed). □

Corollary 4.10.1. A *bijective* mapping f (i.e., into, onto and one-to-one) is a homeomorphism iff $f(\bar{A}_1)) = \overline{f(A_1)}$ $\forall A_1 \subseteq X_1$. Now assume that f is a bijective mapping satisfying the condition. Thus f is continuous by Theorem 4.2 as $f(\bar{A}_1) = \overline{f(A_1)}$. Similarly $f^{-1} = g$ is continuous. Thus f is a homeomorphism.

In other words, for a bijection openness, closedness or continuity of the inverse function are equivalent requirements. Note that this is only true for a bijection. It is the failure of this requirement because of which the projection mapping from $\mathbb{R}^2$ to $\mathbb{R}^1$ can be open without being closed.

We are now in a position to identify more clearly what the topological properties are that are preserved by homeomorphisms. We should then be able to state more precisely what we mean when we say that two topological spaces are equivalent, or for that matter different. This is done in the next section.

4.5 Topological Properties and Homeomorphisms

The first requirement that we must impose on our definition of equivalent topological spaces is that the relationship between them is an *equivalence relation* in the sense of set theory. Before proceeding on to topological equivalence we shall briefly recapitulate what an equivalence relation, $\mathcal{R}$, is. If the objects being discussed are denoted by $x, y, z, \cdots$ then $\mathcal{R}$ is: (a) *reflexive*, i.e., $x\mathcal{R}x$; (b) *symmetric*, i.e., if $x\mathcal{R}y$ then $y\mathcal{R}x$; and (c) *transitive*, i.e., if $x\mathcal{R}y$ and $y\mathcal{R}z$ then $x\mathcal{R}z$. For example, the ordering relationship on $\mathbb{R}$, ">" or on sets "$\supsetneq$", satisfies property (c) but not (a) or (b). Again, "$\leq$" on $\mathbb{R}$ or "$\subseteq$" on sets satisfies (c) and (a) but not (b). The relationship "$=$" on $\mathbb{R}$ or on sets satisfies all three requirements and hence is an equivalence relation. Similarly, in Group Theory, isomorphism is an equivalence relation.

Let us now return from our flights into Algebra to the geometrical world of Topology. Our definition of equivalence was based on homeomorphisms. We thus need to prove that the property of "being homeomorphic to" denoted by "$\cong$" is an equivalence relation. This is done in the following theorem.

Theorem 4.11. *The relation "$\cong$" is an equivalence relation.*

Proof. For property (a) $id : (X, \tau) \to (X, \tau)$, the identity map, is obviously a homeomorphism. Thus $(X, \tau) \cong (X, \tau)$. For property (b) let $(X_1, \tau_1) \cong (X_2, \tau_2)$, i.e., $\exists$ a homeomorphism $f : (X_1, \tau_1) \to (X_2, \tau_2)$. Thus $f^{-1} : (X_2, \tau_2) \to (X_1, \tau_1)$ is also a homeomorphism. Hence $(X_2, \tau_2) \cong (X_1, \tau_1)$. For (c) let $(X_1, \tau_1) \cong (X_2, \tau_2)$ and $(X_2, \tau_2) \cong (X_3, \tau_3)$. Thus $\exists$ homeomorphisms $f : (X_1, \tau_1) \to (X_2, \tau_2)$ and $g : (X_2, \tau_2) \to (X_3, \tau_3)$. So the composition $h = g \circ f$ is a homeomorphism by Theorems 4.7, 4.8 and 4.10 read together. Thus $h : (X_1, \tau_1) \to (X_3, \tau_3)$ is a homeomorphism and hence $(X_1, \tau_1) \cong (X_3, \tau_3)$. $\square$

We may not always want to talk about equivalent spaces but perhaps also of *restrictions* of one space being equivalent to another or of *extensions* being equivalent. To develop the terminology in this regard, consider a continuous injective function $f : (X_1, \tau_1) \xrightarrow{into} (X_2, \tau_2)$. If $f : X_1 \to f(X_1)$ is a homeomorphism we can identify X_1 with its image. Thus we can think of (X_1, τ_1) as a subspace of (X_2, τ_2). In this case (X_1, τ_1) is said to be *topologically embedded* in (X_2, τ_2) and the mapping is said to be a *topological embedding*. Here $(X_1, \tau_1) \not\cong (X_2, \tau_2)$ but $(X_1, \tau_1) \subset (X_2, \tau_2)$ in general. Thus (X_1, τ_1) is equivalent to a *restriction* of (X_2, τ_2) to a subspace. Contrariwise, $(X_2, \tau_2) \supset (X_1, \tau_1)$ in general, which says that (X_2, τ_2) is equivalent to an *extension* of (X_1, τ_1) to a larger space in which it is a subspace. The latter relation is often used to find appropriate extensions of solutions of equations beyond their original domain of definition. We will not go into that use here. However, it is worth while to give some trivial examples to clarify the concepts presented.

Example 1. Take $X_1 = \{a,b\}, \tau_1 = \{\phi, \{a\}, X_1\}; X_2 = \{1,2\}, \tau_2 = \{\phi, \{1\}, X_2\}$. Then $f : (X_1, \tau_1) \to (X_2, \tau_2)$ given by $f(a) = 1, f(b) = 2$ is a homeomorphism and so $(X_1, \tau_1) \cong (X_2, \tau_2)$. Notice that $f(a) = 2, f(b) = 1$ would not be allowable as $f(\{a\})$ would not be open in X_2.

Example 2. Take $X_1 = \{a,b\}, \tau_1 = \{\phi, \{a\}, X_1\}; X_2 = \{1,2,3\}, \tau_2 = \{\phi, \{a\}, X_2\}$. Then $f : (X_1, \tau_1) \to (X_2, \tau_2)$ given by $f(a) = 1, f(b) = 2$ is a topological embedding and so $(X_1, \tau_1) \subset (X_2, \tau_2)$. Equivalently, (X_2, τ_2) is equivalent to an extension of (X_1, τ_1) by making $X_1' = \{a,b,c\}, f^{-1}(3) = c$, so that $f : (X_2, \tau_2) \to (X_1', \tau_1')$ is a homeomorphism.

A property P, possessed by a topological space (X_1, τ_1) which is necessarily possessed by any other topological space $(X_2, \tau_2) \cong (X_1, \tau_1)$ is said to be a *topological property* or a *topological invariant*. Let us look at examples of topological properties.

Theorem 4.12. Second countability is a topological property.

Proof. Consider a homeomorphism $f : (X_1, \tau_1) \to (X_2, \tau_2)$, with X_1 being second countable. Thus X_1 has a countable base for its topology. Let one such base be $\beta_1 = \{B_1, B_2, \cdots, B_n, \cdots\} = \{B_i | i \in \mathbb{N}\}$, where $B_i \in \tau_1$ and $\forall\, U_1 \in \tau_1$ we can write $U_1 = \bigcup_{i \in I} B_i, I \subseteq \mathbb{N}$. Then, as f is a homeomorphism, $\forall B_i, f(B_i) \in \tau_2$. Also, since $X_1 = \bigcup_{i \in \mathbb{N}} B_i$, $X_2 = f(X_1) = f\left(\bigcup_{i \in \mathbb{N}} B_i\right) = \bigcup_{i \in \mathbb{N}} f(B_i)$, i.e., all the images of the elements of the base are open and their union is the whole of the image space. Further, if $V \in \tau_2$, $f^{-1}(V) \in \tau_1$. Thus $f^{-1}(V) = \bigcup_{i \in I} B_i$ and so $f\left(f^{-1}(V)\right) = f\left(\bigcup_{i \in I} B_i\right) = \bigcup_{i \in I} f(B_i)$. Thus $\forall V \in \tau_2$ we can write $V = \bigcup_{i \in I} f(B_i)$. Also $f(B_i)$ is countable. Thus we have constructed $\beta_2 = \{f(B_1), f(B_2), \cdots, f(B_n), \cdots\} = \{f(B_i) | i \in \mathbb{N}\} = f(\beta_1)$, which serves as a countable bases and therefore second countability is a topological property. $\square$

Theorem 4.13. First countability is a topological property.

The proof of this theorem follows the same lines, using a local base. You should try giving the proof in full detail as an exercise.

Theorem 4.14. Separability is a topological property.

Proof. Consider $f : (X_1, \tau_1) \to (X_2, \tau_2)$ s.t. f is a homeomorphism, $X_1 \supseteq A_1$ s.t. $\bar{A}_1 = X_1$ and $|A_1| = \aleph_0$. Clearly, $|f(A_1)| = \aleph_0$. Consider $x_2 = f(x_1)$ for some $x_1 \in X_1$ and $V \in \tau_2$ s.t. $x_2 \in V$. Then $f^{-1}(V) \in \tau_1$ and $x_1 \in f^{-1}(V)$. Since $\bar{A}_1 = X_1, f^{-1}(V)$ contains some $a_n \in A_1$. Therefore $f(a_n) \in f(A_1) \cap V \neq \phi$. Hence $x_2 \in \overline{f(A_1)}$, i.e., $\overline{f(A_1)} \subseteq X_2 \subseteq f(\bar{A}_1)$. Hence $X_2 = \overline{f(A_1)}$. Hence X_2 is also separable. $\square$

4.6 Exercises

1. Let X and Y be two metric space. Show that a constant mapping $f : X \to Y$ is continuous. Is f an open mapping? Is f a closed mapping? Verify!

2. Let $f : X \to Y$ be a continuous mapping of a topological space X into a topological space Y, and let $\beta = \{B_\alpha\}$ be a bases for the topology of X. Show that f is an open mapping iff $f(B_\alpha)$ is open in Y.

3. Let X be a discrete space and Y any topological space. Show that any mapping $f : X \to Y$ is always continuous.

4. Show that a function $f : X \to Y$ is continuous iff for each $B \subseteq Y, f^{-1}(int(B)) \subseteq int(f^{-1}(B))$.

5. Let $X = \bigcup_\alpha X_\alpha$, where each X_α is an open subset of X, and let $f : X \to Y$ be a mapping. If the restriction $f_\alpha = f\big|_{X_\alpha}$ is continuous for each α, show that $f : X \to Y$ is continuous.

6. Show that the mapping $f : \mathbb{R} \to \mathbb{R}$ defined by $f(x) = |x| \ \forall x \in \mathbb{R}$, is continuous. [Hint: Let $x_n \to x$, then since $||x_n| - |x|| \le |x_n - x|$ continuity of f follows.]

7. Show that $(-1, 1)$ and $\mathbb{R}$ (with respect to the usual topology) are homeomorphic. [Hint: Show that $f : \mathbb{R} \to (-1, 1)$ defined by $f(x) = x / (1 + |x|), x \in \mathbb{R}$, is a homeomorphism.]

8. Suppose that $(X, \tau_X) \cong (Y, \tau_Y)$ and (X, τ_X) is metrizable. Show that (Y, τ_Y) is also metrizable. (Thus metrizability is a topological property.)

9. Show that first countability and second countability are both topological invariants.

10. Show that completeness is not a topological invariant.

11. Show that compactness is a topological property.

12. Show that connectedness and path connectedness are both topological invariants.

13. Is separability a topological property? Justify your answer.

14. Show that a continuous open (or closed) bijection is a homeomorphism.

15. Suppose that τ_1 and τ_2 are topologies for a non-empty set X and that $id : (X, \tau_1) \to (X, \tau_2)$ is the identity map. Prove that id is continuous iff $\tau_2 \subset \tau_1$. Under what conditions is id a homeomorphism?

16. Suppose that $f, g : X \to \mathbb{R}$ are continuous. Show that $h : X \to \mathbb{R}$ defined by $h(x) = \max\{f(x), g(x)\}$ is continuous.

17. Let $f : X \to Y$ be a mapping and $A \subset X$. Show that $f\big|_A$ may be continuous even though f is not continuous at any point of A. [Hint: Consider $f : \mathbb{R} \to \mathbb{R}$ defined by

$$f(x) = \begin{cases} 1 & \text{if } x \in \mathbb{Q} \\ 0 & \text{otherwise} \end{cases}$$

Then $f\big|_\mathbb{Q}$ is continuous, however f (without the restriction) is not continuous.]

18. Find mappings that are open but not closed, closed but not open, continuous but not open or closed.

19. Let $\mathbb{R}$ be equipped with the half-open interval topology τ, and let $A = [0, 1)$ with the induced topology. Show that A is homeomorphic to $\mathbb{R}$.

Chapter 5

The Separation Axioms

In the process of generalizing away from more common concepts of space to unusual concepts, it is often unclear how far we should go. There is a very real danger of generalizing to the point of absurdity, where nothing is gained by further generalizations. Thus, when we start extending from spaces on which geometry resembles what we are used to, we do not want to end up with just a structureless set. On the other hand, we do not want to be too severely limited by our prejudices of what is usual. To avoid this problem we can develop gradations of generalization, so that we can choose how far we want to go from our original concepts. To some extent that is what we have done in going from plane geometry to the geometry of surfaces on the one hand and higher dimensions on the other. The next extension, to curved higher dimensional spaces, really requires the use of concepts of metric spaces with the usual metric. Further extensions are provided by taking arbitrary metrics or going to topological spaces, without even requiring a metric. We have left unanswered the question of which topological spaces are non-metrizable. We are now in a position to begin answering this question.

To be able to deal with assigning a metric to a topological space we need to know how well we can separate points of the space from each other. Our geometrical intuition is based on the set of real numbers, and their Cartesian products. For such spaces we can separate any two points fully by putting them in separate open intervals each. However, in a general topological space this is clearly not possible. In the indiscrete topology all elements belong to the unique open set which is the universal set. On the other hand, in the discrete topology we have *total* separation, where each element is its own neighborhood, as it were. The former can be thought of as an extremely closely knit colony, somewhat like a multi-cellular living organism where one part cannot be cut off from the rest without impairing the whole. The latter as a contradiction to Donne's claim that "No man is an island complete unto himself...", a "society" where "Every man is an island complete unto himself", is a set of isolated individuals who see only one distinction; themselves and everybody else. Like in the social context, one would want to find a "happy medium" between them. As there, the choice of "medium" depends on individual requirements.

In this chapter we will consider three possible "happy mediums" to choose from – starting from the bizarre, to the merely exotic and ending with the usual. We will go on to introduce a fourth, which will be more than merely "usual". We will then go on to consider various ways of characterizing these spaces so as to explore the interrelationships between them and study each in its own context. We will thus be able to place more precisely where a particular space lies as

5.1 The Three Separation Axioms

When we talk of separation *axioms* we do not mean that these are *independent* axioms. Rather, each is a step towards a *stronger axiom*. We may choose to stop at each stage, or to go on to the next axiom. As such we have an increasingly stronger structure and less generality. We exclude, in our steps, the indiscrete or the discrete topologies where *no* separation is possible or *total* separation is guaranteed. Our definitions of the extent of separability of the space are also limited by *democratic* requirements in that we will not provide a separate classification where one special element is separated from the rest, but the others are all lumped together — like a King and his subjects. For us, all elements are equally good. Thus a topology on a space $X, \tau = \{\phi, \{a\}, X \setminus \{a\}, X\}$, for some $a \in X$ (which one could call a *monarchic* topology), will not be a special category. Whatever condition we state will be required to apply *universally* to all elements of the set.

The weakest structure that fits our conditions is that $\forall x, y \in X, \exists$ a nbd of one of the points that does not contain the other. For example, if $X = \{a, b, c\}$ and $\tau = \{\phi, \{a\}, \{a, b\}, \{a, c\}, X\}$, we have our requirement met. For $a, b \in X$ we have the choice $U_a = \{a\}$, $U_b = \{a, b\}$ s.t. $b \notin U_a$; for $b, c \in X$ we have the choice $U_b = \{a, b\}$, $U_c = \{a, c\}$ s.t. $b \notin U_c$ and $c \notin U_b$. Generally, we require that either $x \notin U_y$ or $y \notin U_x$ (or of course both). Such a space is called a T_0–space or a *Kolmogorov space*.

A stronger requirement is that $\forall x, y \in X, x \notin U_y$ and $y \notin U_x$. As we shall see later on, the only finite topological space which satisfies this requirement is a discrete space. Thus the only non-trivial examples of such a space are of infinite spaces. We shall shortly look at some such examples. These spaces are already much closer to our usual intuition-based spaces which have to be continuously infinite. They are called T_1–*spaces*.

A still stronger requirement is that $\forall x, y \in X, \exists\ U_x$ and U_y s.t. $U_x \cap U_y = \phi$. Thus, not only do the points in the space lie in separate neighborhoods, even the neighborhoods have no overlap. The points are totally "housed off" from each other. These *are* the usual spaces on which geometry with calculus is used. They are called T_2–spaces. Since this axiom was, coincidentally, stated by Felix Hausdorff, they are called *Hausdorff spaces*. Clearly $\mathbb{Q}$, with the usual topology, is Hausdorff as $\forall x, y \in \mathbb{Q}$, we can define $\epsilon = (y - x)/2$ and $U_x = (x - \epsilon, x + \epsilon)$ so that $U_x \cap U_y = \phi$, because $x + \epsilon = (y + x)/2 = y - \epsilon$. As such $(\mathbb{R}, \tau_u)$ is also obviously Hausdorff.

Let us look at some examples of these three classes of spaces. Remember that every T_2–space is obviously a T_1–space also, but not *vice versa*. Thus a T_2–space is also necessarily a T_0–space, but not *vice versa*. A discrete space is a T_0–, T_1– and a T_2–space while an indiscrete space is not even T_0.

Example 1. Define the topology on $\mathbb{N}$,

$$\tau = \{\phi, A_n = \{n, n+1, \cdots\} | n \in \mathbb{N}\}. \tag{5.1}$$

Consider $i, j \in \mathbb{N}$ s.t. $i < j$. Then $j \in A_i$ but $i \notin A_j$. In fact, for any $n \geq i$ and $n < j, j \in A_n$ but $i \notin A_j$. Hence the space is T_0 but not T_1.

Example 2. Consider (X, τ_c). Then $\forall x, y \in X (x \neq y)$, $X \setminus \{x\}$ and $X \setminus \{y\}$ are open sets containing y and x respectively. Hence X is a T_1-space. However $(X \setminus \{x\}) \cap (X \setminus \{y\}) \neq \phi$. Hence this is not a T_2-space.

Example 3. A set X having 3 or more elements, with the topology defined by

$$\tau = \{\phi, A, A', X\}, \tag{5.2}$$

with any $A \neq \phi, X$, is not a T_0-space.

Example 4. Consider a metric space (X, ρ) and $x \neq y \in X$. Now, let $\rho(x, y) = r$. Choose $\epsilon \leq r/2$ and construct two open spheres $S_\epsilon(x)$ and $S_\epsilon(y)$ centred at x and y respectively. Clearly, not only $x \notin S_\epsilon(y)$ and $y \notin S_\epsilon(x)$, but $S_\epsilon(x) \cap S_\epsilon(y) = \phi$, see Figure 5.1. Hence the space is T_2, and consequently T_0 and T_1.

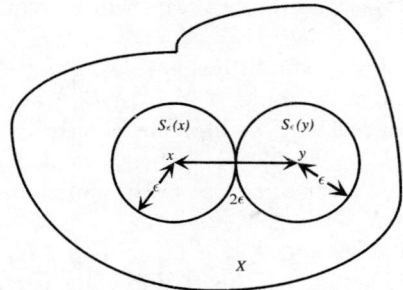

Figure 5.1: For a metric space we can always define a distance between two points x, y of the space. Call it 2ϵ. Now we can construct two spheres, $S_\epsilon(x)$ and $S_\epsilon(y)$ of radius ϵ, and centred at x and y respectively, which do not intersect. Hence the metric space must be a T_2-space.

Example 5. Let $A = [0, 1)$. Then the A-inclusion topology on $\mathbb{R}$ does not provide a T_0-space, as can be easily verified. Again, for any A with $|A| \geq 2$, the A-exclusion topology on $\mathbb{R}$ does not give a T_0-space.

Example 6. Take $X = \mathbb{R}^2$ with the topology generated by the bases

$$\beta = \{(a, b) \times \mathbb{R} | a, b \in \mathbb{R}\}. \tag{5.3}$$

This topology may be thought of as open infinite strips of all thicknesses at all places and is depicted (via the bases) in Figure 5.2. Then $(\mathbb{R}^2, \tau_\beta)$ is not a T_0-space. Similarly, if we put

$$\beta' = \{\mathbb{R} \times (a, b) | a, b \in \mathbb{R}\}, \tag{5.4}$$

whose diagrammatic representation may be obtained by rotating Figure 5.2 through a right angle, does not give a T_0-space. However

$$\overline{\beta} = \{(a, b) \times (a, b) | a, b \in \mathbb{R}\}, \tag{5.5}$$

depicted in Figure 5.3, does give a T_1-space, as may be easily verified.

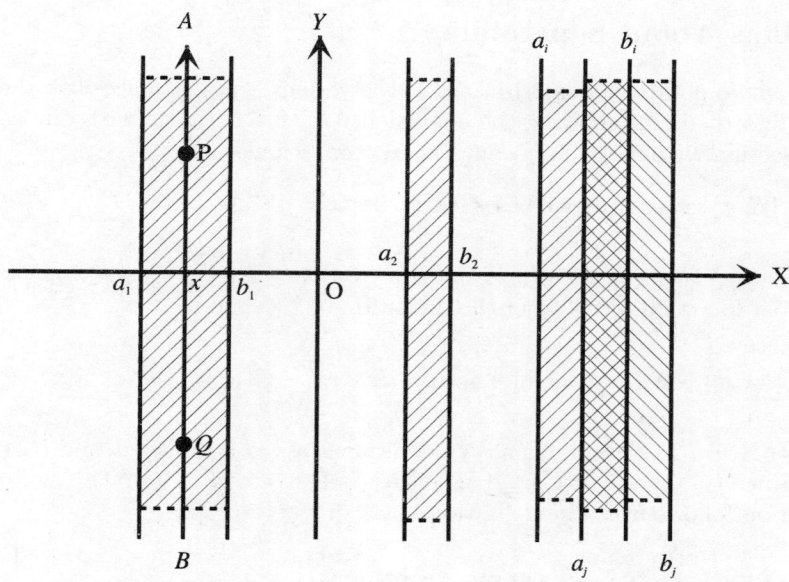

Figure 5.2: The topology generated by taking a bases of $(a,b) \times \mathbb{R}$ may be thought of as infinite strips as shown here. A point $x \in (a_1, b_1)$ will correspond to a "thread" AB in the "ribbon" $(a_1, b_1) \times \mathbb{R}$. Two distinct points on the thread, P and Q, will share common *nbds* as the open set containing the x part will be the same even though $P = (x, y_1), Q = (x, y_2)$. Of course the thickness of the ribbon is arbitrary, e.g. $(a_2, b_2) \times \mathbb{R}$ is another ribbon. Also, the strips must overlap so that $[(a_i, b_i) \times \mathbb{R}] \cap [(a_j, b_j) \times \mathbb{R}] \neq \phi$ for some j for every i.

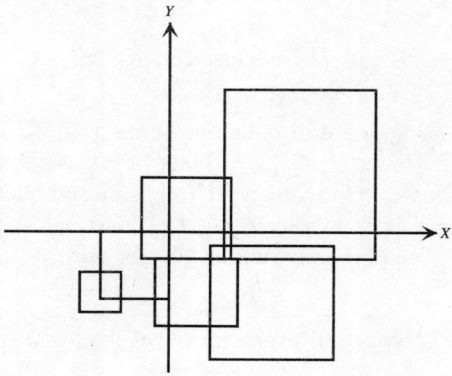

Figure 5.3: The bases generated by $(a,b) \times (a,b)$ of open squares of side $(b-a)$ centred at $\left(\frac{a+b}{2}, \frac{a+b}{2}\right)$. We can fill up the entire plane with such squares.

Example 7. $\mathbb{R}$ with the right ray (or the left ray) topology is not a T_1–space but *is* T_0, as we have for any $a, b \in \mathbb{R}$ if $a < b, b \in U_a$ but $a \notin U_b$ (or $a \in U_b$ but $b \notin U_a$). Similarly, $\mathbb{Z}$ with the left or right ray topology is T_0 but not T_1.

5.2 Theorems About Separability

As before, we want to obtain characterizations of the various types of spaces by other criteria and will, therefore, find theorems showing the equivalence of those criteria with our definitions very useful. In this section we provide many such characterizations.

Theorem 5.1. (X, τ) generated by a bases β is T_0 iff $\forall x \neq y \in X, \exists$ either U_x or $V_y \in \beta$ s.t. $y \notin U_x$ or $x \notin V_y$.

Proof. The proof follows directly from the definition of T_0–spaces. $\square$

Theorem 5.2. (X, τ) is a T_0–space iff $\forall x \neq y \in X, \{x\}' \in \mathfrak{N}(y)$ or $\{y\}' \in \mathfrak{N}(x)$.

Proof. Suppose X is a T_0–space. Then, by definition $\exists$ a nbd U_x of x s.t. $y \notin U_x$ or a nbd U_y of y s.t. $x \notin U_y$. Since $U_x \subseteq \{y\}'$ or $U_y \subseteq \{x\}'$ either $\{x\}'$ is a nbd of y or $\{y\}'$ is a nbd of x. The reverse implication follows from the definition. $\square$

Theorem 5.3. (X, τ) is a T_0–space iff $\mathfrak{N}(x) = \mathfrak{N}(y) \Leftrightarrow x = y \ \forall x, y \in X$.

Proof. This is left as a simple exercise for you. $\square$

Theorem 5.4. (X, τ) is a T_1–space iff $\forall x \neq y \in X, \exists U_x, V_y \in \beta$ s.t. $x \notin V_y$ and $y \notin U_x$.

Proof. The proof follows directly from the definition of the topological bases and the T_1–space. $\square$

Theorem 5.5. (X, τ) is a T_1–space iff $\{x\}$ is closed in $X, \forall x \in X$.

Proof. Assume that X is a T_1–space and consider some $x \in X$. Then, for any $y \neq x \ \exists \ U_y \in \beta$ s.t. $x \notin U_y$ (by Theorem 5.4). Hence $U_y \subseteq \{x\}'$. Thus $\{x\}'$ is a nbd of all points in it. Hence $\{x\}'$ is open and therefore $\{x\}$ is closed. Conversely, if $\{x\}$ is closed $\forall x \in X$ then $\forall y \neq x \ \exists \ \{x\}' \ni y$ which is open and $\{y\}' \ni x$ which is open. Since $x \notin \{x\}'$ and $y \notin \{y\}'$ by definition we have open nbds $\{y\}'$ and $\{x\}'$, of x and y respectively, which do not contain y and x, respectively. Hence (X, τ) is a T_1–space. $\square$

Corollary 5.5.1. (X, τ) is a T_1–space iff every finite set is closed in X.

Proof. It follows automatically from the above theorem. $\square$

Corollary 5.5.2. (X, τ) is a T_1–space iff $\tau_c \subseteq \tau$.

Proof. If (X, τ) is a T_1 space, consider $A \in \tau_c$. Then A' is a finite set, say $A' = \{x_1, \cdots, x_n\}$. By the previous corollary A' is closed in (X, τ). Hence $(A')' = A \in \tau$. This proves the forward implication. Conversely take $\tau \supseteq \tau_c$. Thus if $A \in \tau$, A' is a countable set which is closed. Hence (X, τ) is a T_1–space by the previous corollary. $\square$

Theorem 5.6. If (X, τ) is a T_1-space and x is a limit point of $A \subseteq X$ then every *nbd* of x contains infinitely many distinct points of A.

Proof. If x is a limit point of $A, x \in \bar{A}$, suppose $\exists$ a *nbd* U of x s.t. $U \cap A = \{x_1, \cdots, x_n\}$. Since X is a T_1-space, $\exists$ *nbds* U_i of x s.t. $x_i \notin U_i \; \forall i = 1, \cdots, n$. Let $V = \bigcap_{i=1}^n U_i$. Then V is a *nbd* of x s.t. $V \cap A = \phi$. Thus we have a *nbd* of x which does not intersect A. Hence x cannot be a limit point of A, in contradiction to the earlier assumption. Thus the supposition that $U \cap A$ has a finite number of elements is wrong. Since the argument relies on the *finite* intersection $\bigcap_{i=1}^n U_i$, it will break down if the index set is infinite. Hence we can only have an infinite number of points in $U \cap A$. This completes the proof. □

Theorem 5.7. If (X, τ) has a topological bases β then X is a T_2-space iff $\forall x \neq y \in X \; \exists$ *nbds* η_x and η_y (of x and y respectively) in β s.t. $\eta_x \cap \eta_y = \phi$.

Proof. As with Theorem 5.1 and Theorem 5.4, the theorem follows from the definition of the base and a T_2-space. □

There are many other theorems regarding T_2-spaces which will be stated later. Before proceeding to them we need to demonstrate that the property of being T_0, T_1 or T_2 is a topological property. We also need to look at examples of these spaces and, as usual, examples where the space *fails* to satisfy the criteria. It will be particularly useful to see examples where some theorem is applied.

Theorem 5.8. Being a T_0-space is a topological property.

Proof. Suppose f is a homeomorphism from a T_0-space (X_1, τ_1) to some other space (X_2, τ_2). To show that X_2 is also a T_0-space, consider $x_2 \neq y_2 \in X_2$ which are images under f of $x_1 \neq y_1 \in X_1$. Since (X_1, τ_1) is $T_0, \exists$ either $U_1 \ni x_1$ s.t. $U_1 \in \tau_1$ or $V_1 \ni y_1$ s.t. $V_1 \in \tau_1$, s.t. $y_1 \notin U_1$ or $x_1 \notin V_1$. Hence $f(U_1) = U_2 \ni x_2$ s.t. $U_2 \in \tau_2$ and $f(V_1) = V_2 \ni y_2$ s.t. $V_2 \in \tau_2$ as f is an open mapping. It is obvious that $y_2 \notin U_2$ or $x_2 \notin V_2$. This completes the proof. □

We have only demonstrated that T_0-spaces are invariant under homeomorphisms but they need not be preserved under open or closed mappings only. This fact appears in the following example.

Example 1. Consider the mapping $f : \mathbb{Z} \to \{0, 1\}$, acting on the integers with the right ray topology, given by

$$\left. \begin{aligned} f(n) &= 1 \quad \text{if} \quad n = 2p+1, \\ &= 0 \quad \text{if} \quad n = 2p, \end{aligned} \right\} \; (p \in \mathbb{Z}), \tag{5.6}$$

and endow $\{0, 1\}$ with the indiscrete topology. It is easy to verify that f is an open and closed continuous function. However, $\mathbb{Z}$ with the right ray topology is a T_0-space but $\{0, 1\}$ with the indiscrete topology is not.

Theorem 5.9. Being a T_1–space is a topological property.

Proof. The proof follows the same reasoning as Theorem 5.8, except that here the "or" is replaced by an "and". □

Example 2. The f given in the previous example acting on $\mathbb{Z}$ with the co-finite topology with the image space the same as before is an open continuous mapping from a T_1–space to a space that is not even T_0. Hence it is not necessary that open continuous mappings should preserve T_1–spaces.

Theorem 5.10. Being T_1 is preserved by closed mappings.

Proof. Take f to be a closed mapping from a T_1–space (X_1, τ_1) to some (X_2, τ_2). Since every singleton $\{x\}$ is closed in X_1 and f is a closed mapping, $f(\{x\})$ is closed in X_2. Thus every singleton is closed in X_2 and hence X_2 is a T_1–space by Theorem 5.5. □

Exercise: Define a topology on $\{a,b\}$ so that it is T_0 but not T_1. Do the same for $\{x_1, \cdots, x_n\} (n \geq 2)$.

5.3 Product Topologies and Product Spaces

Given a collection of sets $\{X_i\}$, we can define their Cartesian product. Whereas there is no limitation on the index set in principle, we will limit our discussion to a finite index set to define the *product set* $\prod_{i=1}^{n} X_i$. (As an example keep in mind $\mathbb{R}^2 = \mathbb{R} \times \mathbb{R}$, or more generally $\mathbb{R}^n = \prod_{i=1}^{n} \mathbb{R}$.) Now suppose each of the sets has an associated topology, τ_i. We could ask for the topology which should be associated with the product set. Setting $X = \prod_{i=1}^{n} X_i$, we want the new topological space (X, τ) which should naturally be associated with the product of topological spaces, called the *product space*. Here the topology, τ, called the *product topology*, is denoted by $\tau = \prod_{i=1}^{n} \tau_i$. The question arises as to what we really mean by this product. Do we mean that every element of τ is the Cartesian product of elements of τ_i? This cannot be true in general. For example in $(\mathbb{R}, \tau_u) \times (\mathbb{R}, \tau_u)$ we could take $\phi \times (a, b)$, which would not be meaningful in $(\mathbb{R}^2, \tau_u)$. We would, naturally, expect to get the product topology of two usual topologies to be the usual topology.

Though it may not necessarily be possible to write all products of open sets as open sets, we would certainly require that the product of non-empty open sets in the constituent topologies be an open set in the product topology. In other words we can define a bases

$$\beta = \left\{ U = \prod_{i=1}^{n} U_i | U_i \in \beta_i \right\}, \tag{5.7}$$

for the *product topology*. It may not be easy to use this definition to specify the bases of a product topology. It would be more convenient to use a product of the bases of each constituent topology. That this can be done is guaranteed by the next theorem.

Theorem 5.11. For a finite set of topological spaces $\{(X_i, \tau_i) | i = 1, \cdots, n\}$ with bases β_i generating τ_i, a bases for $X = \prod_{i=1}^{n} X_i$ is

$$\beta = \left\{ B = \prod_{i=1}^{n} B_i | B_i \in \beta_i \right\}. \tag{5.8}$$

Proof. Consider $V \subseteq X$ and any point $\mathbf{x} \in V$. By the above definition (Eq. (5.7)) $\exists U$ s.t. $\mathbf{x} \in U \subseteq V$. Since β_i is a bases for τ_i we can choose $B_i \in \beta_i$ s.t. $\mathbf{x} \in B \subseteq U \subseteq V$. Hence β is a bases for τ by Theorem 5.4. $\square$

Let us look at some examples of product topologies. The most obvious one, referred to earlier, is $\mathbb{R}^n$. For convenience let us restrict our attention to $\mathbb{R}^2$. Then, for $(\mathbb{R}, \tau_u) \times (\mathbb{R}, \tau_u) = (\mathbb{R}^2, \tau_u)$ the bases for the product topology is given by the Cartesian product of the bases of the constituent topologies, $\beta_{1,2} = \{(a,b) | a, b \in \mathbb{R}\}$. Thus

$$\beta = \{(a,b) \times (c,d) | a, b, c, d \in \mathbb{R}\}, \tag{5.9}$$

which are arbitrary open rectangles, see Figure 5.4. In this product space consider the set

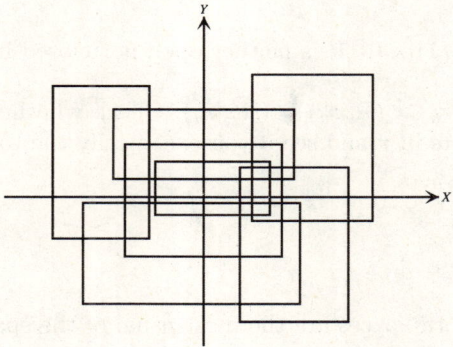

Figure 5.4: We can make an "easier" bases by open rectangles $(a,b) \times (c,d)$, which is a lot more flexible in construction than squares. The sides are $(b-a), (d-c)$ and centres $\left(\frac{a+b}{2}, \frac{c+d}{2}\right)$.

$$A = \{(x_1, x_2) | x_1, x_2 \in \mathbb{R} \ s.t. \ x_1, x_2 \geq 0\}. \tag{5.10}$$

(Here $(x_1, x_2) = \mathbf{x}$ is a point in $\mathbb{R}^2$ and not an interval in $\mathbb{R}$.) Is this set open, closed, both, neither? To answer this question consider

$$A' = \mathbb{R}^2 \setminus A = [(-\infty, 0) \times \mathbb{R}] \cup [\mathbb{R} \times (-\infty, 0)]. \tag{5.11}$$

Since each of the parts is open, their union is open. Hence A' is open and consequently A is closed (see Figure 5.5).

Example 1. The set $M = \{\alpha(1,1) | \alpha \in \mathbb{R}\}$, defined by the usual multiplication of a vector by a scalar, is a closed set. This can be easily demonstrated by the same methods as above, and is left for you to do as an exercise.

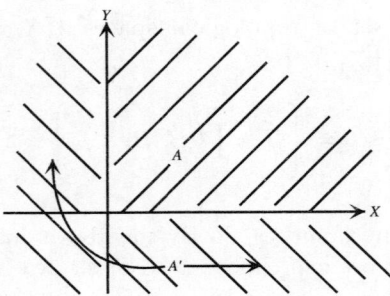

Figure 5.5: A is the "first quadrant" including the coordinate axes adjacent to it. A' is the rest of the plane, and is clearly open. Hence A must be closed. Though it seems "open" to the right and up, remember that the whole plane is both open and closed. Hence, since the boundary is included when a boundary can occur, the set is closed.

Example 2. The circle $\mathbb{S}^1 = \{(x_1, x_2) | x_1^2 + x_2^2 = a^2\}$ is closed in $(\mathbb{R}^2, \tau_u)$. Again, this can be proved (by you) easily. Similarly, the hyperbola $H_1 = \{(x_1, x_2) | x_1 x_2 = a\}$, or equivalently $\{(x_1, x_2) | x_1^2 - x_2^2 = a^2\}$, is closed.

Example 3. The square $[0, 1) \times (0, 1]$ is neither open nor closed in $(\mathbb{R}^2, \tau_u)$ as is obvious.

Example 4. Consider $(\mathbb{R}, \tau_u) \times (\mathbb{R}, \tau_u) = (\mathbb{R}^2, \tau_u)$. Check whether any finite rectangle is ever open in τ. Construct open sets in τ and see if you can specify the topological bases for τ through the constituent topologies.

5.4 Hausdorff Spaces

As mentioned earlier, Hausdorff spaces are the most usual of the spaces discussed in this chapter so far. In fact any space on which usual geometrical methods can be used must be Hausdorff. That this is so is proved in the following theorem.

Theorem 5.12. Every metric space is Hausdorff.

Proof. For $x \neq y \in X$ define $\epsilon = \rho(x, y)/2$. Then $S_\epsilon(x) \cap S_\epsilon(y) = \phi$ while $S_\epsilon(x)$ and $S_\epsilon(y)$ are open sets containing x and y respectively. Since x and y are arbitrary (X, ρ) is a T_2–space. □

Since geometry requires a metric to be defined on the space, that space must be Hausdorff. There is, then, the following statement that is more or less obvious, which needs to be stated explicitly, nevertheless.

Corollary 5.12.1. $(\mathbb{R}^n, \tau_u)$ is a Hausdorff space.

To get a real "feel" for the T_2 restriction we need to see a more usual example of a non-T_2-space than has been seen so far. This may be constructed by a "cut and paste" method. Consider two copies of the Euclidean plane $(\mathbb{R}^2, \tau_u)$, with coordinates $\mathbf{x} = (x_1, x_2)$ and $\mathbf{y} = (y_1, y_2)$ respectively. Consider "open spheres" constructed about $(1, 0)$ in each plane. Thus

$$S_r^1 = \{(x_1, x_2) | (x_1 - 1)^2 + x_2^2 < r^2\}, S_s^2 = \{(y_1, y_2) | (y_1 - 1)^2 + y_2^2 < s^2\}. \tag{5.12}$$

Now cut out the unit open disc from each plane and identify them, using coordinates $\mathbf{z} = (z_1, z_2)$ there, so that

$$D \equiv D^1 \equiv D^2 = \{(z_1, z_2) | z_1^2 + z_2^2 < 1\}. \tag{5.13}$$

Clearly, $(1, 0) \notin D$. Hence we have $(1, 0)_1$ and $(1, 0)_2$ as two distinct points in this space, see Figure 5.6. Their *nbds* are such that they contain S_r^1, S_s^2 for some choices of r and s. Clearly

$$\left.\begin{array}{l} S_r^1 \cap D = \{(z_1, z_2) | z_1^2 + z_2^2 < 1 \text{ and } (z_1 - 1)^2 + z_2^2 < r^2\} \neq \phi, \\ S_s^2 \cap D = \{(z_1, z_2) | z_1^2 + z_2^2 < 1 \text{ and } (z_1 - 1)^2 + z_2^2 < s^2\} \neq \phi. \end{array}\right\}. \tag{5.14}$$

Hence, putting $t = \min\{r, s\}$, which is positive,

$$S_r^1 \cap S_s^2 = \{(z_1, z_2) | z_1^2 + z_2^2 < 1 \text{ and } (z_1 - 1)^2 + z_2^2 < t^2\} \neq \phi. \tag{5.15}$$

Hence the space is non-Hausdorff.

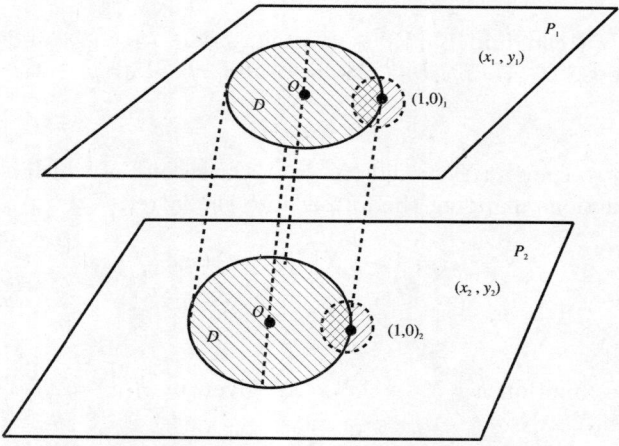

Figure 5.6: A non-Hausdorff space $\mathbb{P}_1 \cup \mathbb{P}_2$ in which the open disc $D_1 \subseteq \mathbb{P}_1$ and $D_2 \subseteq \mathbb{P}_2$ is identified (pasted together) to give the single open disc D. Now $(1, 0)_1 \in \mathbb{P}_1$ and $(1, 0)_2 \in \mathbb{P}_2$. Any "open spheres" centred about these points must overlap inside D in the cross-hatched area. The smaller of the two "open spheres" intersecting D will give the intersection of the *nbds* of $(1, 0)_1$ and $(1, 0)_2$.

How is it that this is not a T_2–space despite Theorem 5.12? After all, it certainly seems like an $(\mathbb{R}^2, \tau_u)$, which should be T_2, by the Corollary 5.12.1. The answer is that it is a metric space everywhere except on the unit circle in both planes. Consider $\mathbf{x}$ and $\mathbf{y}$ in the two planes such that $x_i = y_i$ and $x_1^2 + x_2^2 = 1$. In the entire space, taken as a whole, we can construct $\rho(\mathbf{x}, \mathbf{y})$ as the infimum of all path lengths joining $\mathbf{x}$ to $\mathbf{y}$. This is clearly zero. However, $\mathbf{x}$ and $\mathbf{y}$ are distinct points (despite the fact that they have the same numerical values of coordinates), as one lies in one plane and the other in the other plane. Thus $\rho(\mathbf{x},\mathbf{y}) = 0$ with $\mathbf{x} \neq \mathbf{y}$. Thus no metric can be defined for points on the unit circle.

The above example brings out what the T_2 condition entails. It would be a useful exercise for you to construct a simpler example, analogous to the above, using two $(\mathbb{R}, \tau_u)$ copies. Again, you could look for examples in the space of all functions, constructed in a way similar to the above example.

Let us look at other characterizations of T_2–spaces that can often prove useful.

Theorem 5.13. (X, τ) is a T_2–space iff the *diagonal*,

$$\Delta := \{(x,x)|x \in X\} \tag{5.16}$$

(where ":=" stands for "is defined to be") is closed in $X \times X$.

Proof. Assume (X, τ) is a T_2–space. Then, clearly so is $(X^2, \tau^2) := (X, \tau) \times (X, \tau)$. Consider $M = X^2 \setminus \Delta$ and some $(x,y) \in M$ s.t. $x \neq y$. Since (X, τ) is a T_2–space $\exists\, U, V \in \tau$ s.t. $U \cap V = \phi$ and $U \in \mathfrak{N}(x), V \in \mathfrak{N}(y)$ and $(x,y) \in U \times V \subseteq M$. Thus $M = int(M)$ and so $M \in \tau^2$. Hence $\Delta = M'$ is closed. Now assume that Δ is closed and $x \neq y \in X$. Thus $M = \Delta'$ is open and contains (x, y). Since

$$\beta = \{U \times V | U, V \in \tau\} \tag{5.17}$$

form a bases for τ^2, we can find $U_0, V_0 \in \tau$ s.t. $(x,y) \in U_0 \times V_0$ and $(U_0 \times V_0) \cap \Delta = \phi$. Moreover, we can choose U_0, V_0 s.t. $U_0 \cap V_0 = \phi$ and $x \in U_0$, $y \in V_0$ as if $z \in U_0, V_0$ then $(z,z) \in (U_0 \times V_0) \cap \Delta$. $\square$

Theorem 5.14. For two spaces (X, τ) and (X^1, τ^1), the second of which is Hausdorff, if f and g are two continuous functions mapping the former into the latter,

$$A = \{x \in X | f(x) = g(x)\} \tag{5.18}$$

is closed in (X, τ).

Proof. Construct the function $h : X \to X^1 \times X^1$ given by $h(x) = (f(x), g(x))$. Thus we can construct h^{-1} on $X^1 \times X^1$. Now

$$h^{-1}(U \times V) = f^{-1}(U) \cap g^{-1}(V). \tag{5.19}$$

You can convince yourself of the validity of this statement by applying h to both sides of Eq. (5.19). Now if $U \times V$ is open in $X^1 \times X^1$, $h^{-1}(U \times V)$ must be open in X_1 since $f^{-1}(U)$ and $g^{-1}(V)$ are open in X_1. Clearly, $A = h^{-1}(\Delta)$, where Δ is the diagonal in $X^1 \times X^1$. Since X^1 has been taken to be Hausdorff, by the previous Theorem 5.13, Δ is closed in $X^1 \times X^1$ and thus its pre-image, A, must be closed in X. $\square$

Corollary 5.14.1. If two continuous functions, f and g, map a topological space (X, τ) to a Hausdorff space (X^1, τ^1) so that $f(x) = g(x)\ \forall x$ in a dense subset, A, of X then the two functions must be identical.

Proof. Since A, given by Eq. (5.18), is dense in X_1 and closed in X, $f(x) = g(x)\ \forall x \in X_1$. $\square$

Corollary 5.14.2. The *graph*, G_f, of x given by

$$G_f = \{(x, f(x)) | x \in X\}, \qquad (5.20)$$

of a continuous function, f, mapping a topological space into a Hausdorff space, is closed in $X \times X^1$.

Proof. Define the functions $\varphi, \psi : X \times X^1 \to X^1$ by $\varphi(x, x^1) = x^1$ and $\psi(x, x^1) = f(x)$. Thus φ is a projection mapping, which is continuous, and $\psi = f \circ p_x$, where f is the given continuous mapping and p_x is the projection mapping along X, which is also continuous. Hence ψ is continuous. The above theorem tells us that

$$A = \{(x, x^1) | (x, x^1) \in X \times X^1, \ \varphi(x, x^1) = \psi(x, x^1)\}, \qquad (5.21)$$

is closed in $X \times X^1$. It is clear that A given by Eq. (5.21) is simply another way of writing G_f given by Eq. (5.20). Hence G_f is closed in $X \times X^1$. □

We have seen the significance of T_2–spaces. Now let us look at some alternative ways to characterize them.

Theorem 5.15. (X, τ) is a T_2–space iff $\forall x \neq y \in X \ \exists$ a continuous function $f : X \to X^1$, where (X^1, τ^1) is a T_2–space, s.t. $f(x) \neq f(y)$.

Proof. If (X, τ) is Hausdorff, choosing $(X^1, \tau^1) = (X, \tau)$ and $f(x) = x$, obviously $f(x) \neq f(y) \ \forall x \neq y \in X$. Conversely, if $f : X \to X^1$ is a continuous function such that (X^1, τ^1) is Hausdorff and $f(x) \neq f(y)$, choose $U, V \in \tau^1$ s.t. $U \cap V = \phi$ and $f(x) \in U, f(y) \in V$. Then $f^{-1}(U) \cap f^{-1}(V) = \phi$ and $x \in f^{-1}(U), y \in f^{-1}(V)$. Hence X will be Hausdorff. □

Theorem 5.16. All subspaces of a Hausdorff space are Hausdorff.

Proof. Consider a subspace A of a Hausdorff space X and $x_1 \neq x_2 \in A$. Since X is Hausdorff, we can find $U_1, U_2 \in \tau$ s.t. $x_1 \in U_1, x_2 \in U_2$ and $U_1 \cap U_2 = \phi$. Then $V_1 = A \cap U_1$ and $V_2 = A \cap U_2$ are disjoint open sets in A containing x_1 and x_2 respectively. Hence A is a Hausdorff space. □

Theorem 5.17. An open bijective mapping from a Hausdorff space gives a Hausdorff space.

Proof. Consider an open bijective mapping $f : X \to X^1$ and $x_1^1 \neq x_2^1 \in X^1$. Then $\exists x_1, x_2 \in X$ s.t. $x_1 \neq x_2$ and $f^{-1}(x_1^1) = x_1, f^{-1}(x_2^1) = x_2$. Since X is Hausdorff $\exists U_1, U_2 \in \tau$ s.t. $U_1 \cap U_2 = \phi$ and $x_1 \in U_1, x_2 \in U_2$. Hence $f(U_1) \cap f(U_2) = \phi$ and $f(U_1), f(U_2) \in \tau^1, x_1^1 \in f(U_1), x_2^1 \in f(U_2)$, as f is an open mapping. Hence X^1 is Hausdorff. □

Corollary 5.17.1. The T_2–property is topological.

Proof. Since every homeomorphism is a bijective open mapping, if $X \cong X^1$, i.e., the spaces have the same topological properties, there will be a bijective open mapping from one to the other. By the above theorem, then, if one is Hausdorff so is the other. □

It would be a useful exercise for you to prove that a one to one continuous function maps a Hausdorff space to a Hausdorff space. One could also ponder on the question of whether a homeomorphism could map a non-Hausdorff space to a Hausdorff space and similarly whether an open bijective mapping could map a non-Hausdorff space to a Hausdorff space, etc.

5.5 T_3-Spaces

We now come to a separation axiom that is neither a generalization nor a restriction of the previous axioms, but an extension in a different way. We drop our requirement of "democracy" for individual elements and make our conditions of a T_2–space apply to an element, $x \in X$, and a non-empty closed set, $F \subseteq X$, which does not contain x. In other words $\forall\, F \neq \phi$ s.t. $F' \in \tau$ and $x \notin F\ \exists\ \eta_1 \in \mathfrak{N}(x)$ and $\eta_2 \in \mathfrak{N}(F)$ s.t. $U \cap V = \phi$. This condition is depicted diagrammatically in Figure 5.7.

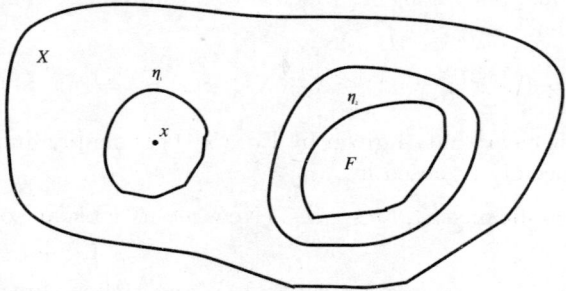

Figure 5.7: The T_3 condition. For any $F \subseteq X$ which is closed and $x \notin F$, we can construct disjoint nbds η_1 and η_2 respectively. The Hausdorff condition need not apply within F unless it can be broken into smaller closed sets.

The condition is *more restrictive* than the T_2 condition because it applies to any *closed set*, F, and not just an *element*, y. However it is *less restrictive* because the Hausdorff condition need not apply within the closed set F. If F contains proper closed subsets it would, but if it has no proper closed subset we have not even required the T_0–condition. To appreciate the distinctions it is worth while to consider some examples of spaces that are, and that are not T_3.

Example 1. Every discrete space is T_3 as every subset is both open and closed in it. Hence $\forall F \subseteq X$ and $x \notin F$, we can take $\eta_1 = \{x\}$ and $\eta_2 = F$ so that $\eta_1 \cap \eta_2 = \phi$. In an indiscrete space there is only one non-empty closed set, namely X. Hence $\nexists\ x$ s.t. $x \notin F = X$. Thus the indiscrete space is obviously not a T_3–space.

Example 2. $X = \{a, b, c\}, \tau_1 = \{\phi, \{a\}, X\}$. If we take $x = a$ we have $F = \{b, c\}$ or X. Now $X \ni x = a$. Thus we must take $F = \{b, c\}$. The only nbd of F is X. Taking $\eta_1 = \{a\}, \eta_2 = X$, we see that $\eta_1 \cap \eta_2 = \{a\} \neq \phi$. Hence the space is *not* T_3. If, instead, we had $\tau_2 = \{\phi, \{a\}, \{b, c\}, X\}$ we see that for $x = a, F = \{b, c\}, \eta_1 = \{a\}, \eta_2 = \{b, c\}$ *are* the respective *nbds* and $\eta_1 \cap \eta_2 = \phi$. Similarly, for $x = b$ or c we have $F = \{a\}$ and $\eta_1 = \{b, c\}, \eta_2 = \{a\}$ which do satisfy the T_3 condition as $\eta_1 \cap \eta_2 = \phi$. Hence this space is T_3. The above example can be generalized for any X and non-null $A \subsetneq X$. With $\tau_1 = \{\phi, A, X\}$ we do not have a T_3–space, as for any $x \in A, F = A'$ is the only non-trivial closed set and $\mathfrak{N}(F) = X = \eta_2$ is unique. Hence $\eta_1 = A$ gives $\eta_1 \cap \eta_2 = A \neq \phi$. However, with $\tau_2 = \{\phi, A, A', X\}$ we do have a T_3–space as, $\forall x \in A$, $\eta_1 = A$ and $F = A' = \eta_2$. Thus $U \cap V = \phi$. Similarly, $\forall x \in A', F = A = \eta_2$ and $\eta_1 = A'$ still gives $\eta_1 \cap \eta_2 = \phi$. This space is not a T_0–space if either $|A|$ or $|A'| \geq 2$. For example, if $A = \{a\}, A' = \{b, c\}\ \nexists$ nbds of b and c which do not contain both elements. Thus we have an example of a T_3-space that is not T_0.

The Separation Axioms

Example 3. Consider $(\mathbb{R}, \tau_c)$. Take any $x \in \mathbb{R}$ and any $F = \{f_1, \cdots, f_n\}$ s.t. $f_i \neq x$, $\forall i = 1, \cdots, n$. Now an open set containing F (which is a closed set, by definition) is typically $G' = V$, where $G = \{g_1, \cdots, g_m\}$ s.t. $g_a \neq f_i$ $\forall a = 1, \cdots, m$, $i = 1, \cdots n$. Also, an open set containing x is $U = A'$, where $A = \{a_1, \cdots, a_p\}$ s.t. $a_\ell \neq x$ $\forall \ell = 1, \cdots, p$. Clearly, $\exists \epsilon > 0$ s.t. $(x - \epsilon, x + \epsilon) \subseteq U, V$. Hence $U \cap V \supseteq (x - \epsilon, x + \epsilon) \neq \phi$. Hence this is not a T_3-space, but is a T_1-space (as seen earlier) and hence a T_0-space.

Example 4. With $X = \mathbb{R}$ define $A = \{1/n | n \in \mathbb{N}\}$. Then the collection of sets

$$\beta = \{U \cup (V \setminus A) | U = (a, b), \ V = (c, d); \ a < b, \ c < d \ \in \mathbb{R}\}, \tag{5.22}$$

serves as a bases for a topology τ. Now

$$\beta_u = \{U | U = (a, b), \ a < b \ \in \mathbb{R}\} \tag{5.23}$$

is a bases for τ_u. Choosing $d < 0$ we see that $\beta \supseteq \beta_u$ and hence $\tau \supseteq \tau_u$. However, $\tau \neq \tau_u$ as $V \setminus A$ can be a union of a semi-open interval $(c, 0]$ with infinitely many open intervals $(1/n, 1/(n-1))$ for $n > 1 + 1/d$ with $d \leq 1$. Hence $\tau_u \subseteq \tau$. Since $(\mathbb{R}, \tau_u)$ is Hausdorff $(\mathbb{R}, \tau)$ must be Hausdorff, i.e., T_2. However, this space is *not* T_3. That this is so is seen by taking $x = 0$ and $0 \notin A = F$. Now, for any *nbd* of 0, $U = (-\epsilon, \delta)$ we have

$$(-\epsilon, \delta) \cap A = \{1/n | n > 1/\delta\} \neq \phi. \tag{5.24}$$

Hence the space is not T_3. Thus we have an example of a space that is T_2 but is not T_3.

From the above examples we see that this new condition is an *extension* of the T_2 concept without being a "generalization" in the sense of being applicable to a wider range of spaces, or in the sense of being a more general requirement. It would be instructive for you to ponder over the question whether there is a meaningful further extension to a "T_4- space" in which we require that $\forall$ closed $F, G \subseteq X$ s.t. $F \cap G = \phi$, $\exists$ nbds η_1 and η_2 of F and G respectively, such that $\eta_1 \cap \eta_2 = \phi$. If so can you construct a T_3-space that is not T_4? Can you construct a T_4-space that is not T_3? How is it related to T_0, T_1, T_2?

5.6 Regular Spaces

Some authors use the term "regular" as synonymous with "T_3". We will *not* use it as such but will follow those authors who define a *regular space* to be a T_3 space which is also T_1. Of our previous examples, we saw that a discrete space is regular as it is T_2 (and hence T_1) and also T_3. None of the other spaces mentioned was regular. An indiscrete space is neither T_3 nor T_0 (leave alone T_1). In Example 2, (X, τ_1) is neither T_3 nor T_0 and (X, τ_2) is T_3 but not T_0, for $|X| \geq 3$. In Example 3, $(\mathbb{R}, \tau_c)$ is T_1 but not T_3 and in Example 4, $(\mathbb{R}, \tau)$ is T_2 but not T_3. Is there, then, any non-discrete regular space?

The simplest example of such a space is $(\mathbb{R}, \tau_u)$! To see this consider any closed $F \subseteq \mathbb{R}$, say $F = [a, b]$ and any $x \notin [a, b]$. Now either $x < a$ or $x > b$. In the former case define $r = (a - x)/2$ so that $x + r = a - r$ and in the latter $r = (x - b)/2$ so that $x - r = b + r$. Then define $\eta_1 \in \mathfrak{N}(x), \eta_2 \in \mathfrak{N}(F)$ by

$$\eta_1 = (x - r, x + r), \quad \eta_2 = (a - r, b + r), \tag{5.25}$$

so that $\eta_1 \cap \eta_2 = \phi$. Thus $(\mathbb{R}, \tau_u)$ is T_3. We know that it is Hausdorff (and hence T_1). Thus it *is* regular.

Another example is the *Fort space* (X, τ_f), where the *Fort topology*, τ_f, is defined for a given $x_0 \in X$, by

$$\tau_f = \{\phi, A \subsetneq X, X | |A'| = n \in \mathbb{N} \text{ or } x_0 \in A'\}, \tag{5.26}$$

i.e., a set is open if its complement is finite or if it does not contain x_0. Clearly, if X is finite A' is finite $\forall A \subseteq X$. Hence we would have a discrete topology. For a non-trivial Fort space X must be infinite. For example, we could have $X = \mathbb{R}$.

Clearly every singleton $\{x\}$, for $x \neq x_0$ is open as $\{x\}' \ni x_0$. Also $\{x\}$ is closed as $(\{x\}')' = \{x\}$ is finite. Thus, for any $x, y \neq x_0$ we can take $\eta_1 = \{x\}, \eta_2 = \{y\}$ to obtain $\eta_1 \cap \eta_2 = \phi$ for $x \neq y$. If $x = x_0 \neq y$, take $\eta_2 = \{y\}$ and $\eta_2 = \{y\}'$ to get $\eta_1 \cap \eta_2 = \phi$. Hence this is a T_2-space. Now for any closed F and $x \neq x_0$ s.t. $x \notin F, \eta_2 = \{x\}'$ is a *nbd* of F and $\eta_1 = \{x\}$ a *nbd* of x s.t. $\eta_1 \cap \eta_2 = \phi$. If $x = x_0$ then $x_0 \notin F$, so $x_0 \in F'$ and hence F' is closed. Thus F is also open. Thus $\eta_1 = F'$ and $\eta_2 = F$ s.t. $\eta_1 \cap \eta_2 = \phi$. Hence the Fort space is T_3. Hence it *is* regular.

We are now in a position to present an alternate characterization of regularity.

Theorem 5.18. A T_1-space (X, τ) is regular iff each point $x \in X$ has a local bases of closed *nbds* (i.e., $\exists$ an open set $U \ni x$ s.t. $U \subseteq F$, where $F' \in \tau$).

Proof. Open *nbds* are anyhow guaranteed. The crucial point is the existence of *closed nbds*. Let η_1 be an open *nbd* of an arbitrary point $x \in X$. To prove the necessity of the condition put $F = \eta_1'$. Since X is regular $\exists$ open *nbds* η_2 and η_3 of x and F respectively (since $\overline{F} = F$) s.t. $\eta_2 \cap \eta_3 = \phi$. Thus $\eta_2 \subseteq \eta_3'$ and so $\overline{\eta_2} \subseteq \overline{\eta_3'} = \eta_3' \subseteq F' = \eta_1$ (as η_3' is closed). Hence the condition is proved. To prove sufficiency suppose that the condition holds and consider any closed F and $x \notin F$. Put $\eta_1 = F'$ and let η_2 be any closed *nbd* of x contained in η_1. Then $\eta_3 = \eta_2'$ is obviously a *nbd* of F s.t. $\eta_3 \cap \eta_2 = \phi$. Hence (X, τ) is regular. This proves the theorem. $\square$

We now proceed to some applications of this theorem about regular spaces.

5.7 Theorems About Regular Spaces

As before, we will provide various theorems about regular spaces for completeness and provide other characterizations.

Theorem 5.19. Every metric space, (X, ρ), is regular.

Proof. Consider any F closed in the topology generated by ρ and $x \notin F$. We define the distance between x and F in the same way as the distance from a point to a plane (see Figure 5.8), namely as the least bound on the distances between the given point and all possible points on the plane. Thus

$$\rho(x, F) = \inf_{y \in F} \rho(x, y) = 2r \text{ (say)} > 0. \tag{5.27}$$

We can construct an open sphere $S_r(x)$ and an open set

$$U = \bigcup_{y \in F} S_r(y), \tag{5.28}$$

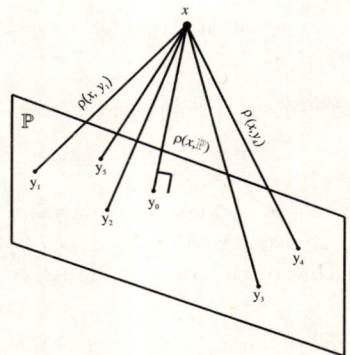

Figure 5.8: The distance from a point x to a set $\mathbb{P}$ may be understood by analogy with the distance from a point to a plane. We can construct the set of all distances from x to points of $\mathbb{P}$ and select the "least" of them. Since we can have a situation like $x = 0$ and $\mathbb{P} = (0,1)$ we may need to use the *infimum* (i.e., the greatest lower bound) rather than simply the minimum. Thus, here $\rho(0, \mathbb{P}) = 0$.

which is obviously an open *nbd* of F. Clearly this is an extension of the proof of the regularity of $(\mathbb{R}, \tau_u)$. To prove that $S_r(x) \cap U = \phi$, suppose $\exists\, z \in S_r(x) \cap U$. Then $\exists\, y_0 \in F$ s.t. $\rho(z, y_0) < r$. Thus by the triangular inequality, we have

$$\rho(z, y_0) \leq \rho(x, z) + \rho(z, y_0) < r + r = 2r, \tag{5.29}$$

which contradicts our definition of r. Hence $\nexists\, z \in S_r(x) \cap U$ and so $S_r(x) \cap U = \phi$. Thus (X, ρ) is T_3. Since every metric space is Hausdorff (and hence T_1) it is regular. □

Theorem 5.20. A subspace of a regular space is regular (i.e., regularity is an inherited property).

Proof. Consider a subspace $X_2 \subseteq X_1$ where X_1 is given to be regular. If $x \neq y \in X_2$ they also belong to X_1. Since X_1 is T_1 $\exists\, U \in \mathfrak{N}(x)$ and $V \in \mathfrak{N}(y)$ s.t. $U, V \in \tau_1$. Then $U \cap X_2$ is an open *nbd* of x and $V \cap X_2$ an open *nbd* of y s.t. $y \notin U \cap X_2$ and $x \notin U \cap X_2$. Hence X_2 is also T_1. Now consider $x \in X_2$ and $F \subseteq X_2$ s.t. $F = \overline{F}$, where $\overline{F}$ denotes closure in X_2 and $x \notin \overline{F}$. Denoting closure in X_1 by $\overline{F}'$, $\overline{F}' \cap X_2 = \overline{F} = F$. Since $x \in X_2$ and $x \notin F, x \notin \overline{F}'$. Since X_1 is regular we can find $U, V \in \tau_1$ s.t. $x \in U, F \subseteq V$ and $U \cap V = \phi$. Thus $U \cap X_2$ and $V \cap X_2$ are open sets in X_2 containing x and F respectively. Hence X_2 is regular. □

Theorem 5.21. Regularity is a topological property.

Proof. To show that regularity is a topological property we must demonstrate that a homeomorphism, f, maps a regular space, X_1, to a space X_2 which is regular. Consider some $F = \overline{F} \subseteq X_2$ s.t. $F \neq \phi$ and some $x_2 \in X_2$ s.t. $x_2 \notin F$. Now, since f is a closed mapping, $f^{-1}(F) = G = \overline{G} \subseteq X_1$ and $f^{-1}(x_2) = x_1 \notin G$. Since X_1 is regular we can choose $U, V \in \tau_1$ s.t. $x_1 \in U, G \subseteq V$. Since f is also an open mapping, $f(U), f(V) \in \tau_2$ s.t. $f(U) \cap f(V) = \phi$ and $x_2 \in f(U), F \subseteq f(V)$. Hence X_2 is regular. □

In Theorem 5.20 we saw that every point of a regular space has a *closed nbd*. We now proceed to prove the converse of that theorem.

Theorem 5.22. (X, τ) is regular if every point of X has a closed *nbd* which is regular as a subspace.

Proof. Let F be any closed *nbd* of $x \in X$, which is regular as a subspace of X. Since $F = \overline{F} \subseteq X$ every member of the local bases β_x is also closed in X. Hence β_x is not only a local bases in F but also a local bases at x in X, consisting of closed *nbds*. Thus, by Theorem 5.18, X is regular. □

Theorem 5.23. A finite product of regular (or T_3–) spaces is also regular (or T_3).

Proof. Consider $(X, \tau) = \prod_{i=1}^{n}(X_i, \tau_i), F = \overline{F} \subseteq X$ and $\mathbf{x} = (x_1, \cdots, x_n) \in X$ s.t. $\mathbf{x} \notin F$. Thus $\exists\, U_i \subseteq X_i$ s.t. $x_i \in U_i \in \tau_i$ and $U \cap F = \phi$, where $U = \prod_{i=1}^{n} U_i$. Hence $F \subseteq U'$. Since each X_i is regular $\exists\, V_i, W_i \subseteq X_i$ s.t. V_i is a *nbd* of x_i and W_i is a *nbd* of U'_i and $V_i \cap W_i = \phi$. Therefore $V = \prod_{i=1}^{n} V_i$ is a *nbd* of $\mathbf{x}$ and $W = \prod_{i=1}^{n} W_i$ of F s.t. $V \cap W = \phi$. Hence (X, τ) is regular. □

Recall that (X, τ) is said to be metrizable if there exists a metric, ρ, on X which generates τ. It can be proved (and you are encouraged to do so) that metrizability is a topological property. Further, subspaces and finite products of metrizable spaces are also metrizable as the topologies of the subspaces and products of metric spaces are the subspace and product topologies respectively.

"Which spaces are metrizable?" remained an open question till the 1950's. However, the crucial step in resolving this question came with *Urysohn's Metrization Theorem*, stated in 1925 by Pavel Urysohn, which showed that regularity (and second countability) were the key conditions. We shall give the statement of the theorem here but not its proof (as that is not at the level of the rest of this book, see [Munkres, 2000] for a standard proof). However, for completeness, we provide the proof in the Appendix.

Theorem 5.24 (Urysohn's Metrization Theorem). A topological space (X, τ) is separable and metrizable iff it is regular and second countable.

This theorem brought out the importance of regularity as a topological property, since metrizability of a space means that it can be treated geometrically.

5.8 Normal Spaces

So far, we have focused on a number of properties, usually known as "The Separation Axioms", other than normality. Note that the separation axioms basically characterize the degree to which points or sets may be kept apart by open sets. An inordinate amount of mathematical energy has been expended in defining separation axioms. We now formulate the *normality axiom*.

A topological space X is said to be a T_4–*space* if for each pair A, B of disjoint closed sets of X there exist disjoint open sets U and V of X containing A and B, respectively.

A topological space is called a *normal space* if it is both T_1 and T_4. It is obvious that every normal space is also regular, however the converse may not be true in general. We illustrate this fact by the following examples.

Example 1. Let X be an indiscrete space consisting of not less than two points. Then X is a T_4–space but not a T_2– space (or even a T_1–space). Note that an indiscrete singleton space is normal.

Example 2. Every discrete space is normal.

Example 3. Let X be any infinite set (for definiteness think of the real line) endowed with the co-finite topology. Then X is not a T_4–space, and hence not a normal space because in such a topology any two non-empty open sets must intersect. Note X is a T_1–space but not a T_2–space.

Example 4. Let $X = \mathbb{R}$ with the p–exclusion topology. Then X is a T_4–space, because if A and B are disjoint closed sets of $\mathbb{R}$ then at least one of them must not contain p. Suppose $p \notin A$. Then A is open by definition of the p–exclusion topology because it excludes point p. Also A' is closed since A is open (by our assumption). Thus A and A' are disjoint neighbourhoods of A and B respectively. Hence X with the p–exclusion topology is a T_4–space. However, we remark that X is not a normal space because it is not a T_1–space.

Example 5. Consider $X = \mathbb{R}$ with the Fort topology (i.e., the minimal topology generated by the p–exclusion topology together with the co-finite topology). Then X is a T_4–space. For, if A and B are any non-empty closed sets then, by the same arguments as given in Example 4, either $p \notin A$ or $p \notin B$. Suppose $p \notin A$ then A is open. Also A' is open because A is closed and hence A, A' are *nbds* of A and B respectively. Note that X is a normal space since it is also T_1.

The following result ensures that every metric space is a normal space and thus provides a class of normal spaces.

Theorem 5.25. *Every metric space (X, ρ) is normal.*

Proof. Let A and B be disjoint non-empty closed subsets of X. Since B is closed, for each $x \in A$ we have
$$\rho(x, B) = \inf_{y \in B} \rho(x, y) = \rho_x > 0. \tag{5.30}$$
Define $U = \bigcup_{x \in A} S_{\rho_x/2}(x)$, where $S_{\rho_x/2}(x)$ is the open sphere with radius $\rho_x/2$ and center at the point x. Clearly, U is an open set containing A. Similarly, we construct $V = \bigcup_{y \in B} S_{\rho_y/2}(y)$ where $\rho_y = \rho(y, A) > 0$, which is also open and contains B. We assert that $U \cap V = \phi$. For if, $z \in U \cap V$, then $z \in S_{\rho_{x_0}/2}(x_0) \cap S_{\rho_{y_0}/2}(y_0)$ for some $x_0 \in A$ and some $y_0 \in B$. By virtue of the triangular inequality
$$\rho(x_0, y_0) \leq \rho(x_0, z) + \rho(z, y_0) < \frac{\rho_{x_0}}{2} + \frac{\rho_{y_0}}{2}. \tag{5.31}$$
If $\rho_{x_0} \leq \rho_{y_0}$ then $\rho(x_0, y_0) < \rho_{y_0}$, so that the open sphere $S_{\rho_{y_0}}(y_0)$ contains the point x_0. If $\rho_{y_0} \leq \rho_{x_0}$, then $\rho(x_0, y_0) < \rho_{x_0}$, so that the open sphere $S_{\rho_{x_0}}(x_0)$ contains the point y_0. In each case, it is impossible. Hence $U \cap V = \phi$. □

We now give several characterizations which are frequently useful in proving normality of certain spaces.

Theorem 5.26. *Let X be a T_1–space. Then X is normal iff given any nonempty closed subset A and its open nbd $U, \exists$ an open set $V \supseteq A$ s.t. $\overline{V} \subseteq U$.*

Proof. First we prove necessity. Assume that X is a normal space and let A be any non-empty closed set and U its open *nbd*. Put $B = U'$. Then A and B are disjoint closed sets. Therefore, by assumption, $\exists$ open *nbds* V and W of A and B respectively such that $V \cap W = \phi$. Hence $V \subseteq W'$ and $\overline{V} \subseteq \overline{W'} = W' \subseteq B' = U$.

Now we prove sufficiency. Let A and B be non-empty disjoint closed subsets of X. Put $U = B'$, then U is an open set containing A. Therefore, by hypothesis, $\exists$ an open set V containing A s.t. $\overline{V} \subseteq U$. Put $W = (\overline{V})'$. Then W is an open set containing B and $V \cap W = \phi$. This completes the proof. $\square$

Theorem 5.27. A T_1–space X is normal iff given disjoint non-empty closed subsets A and B, $\exists$ an open *nbd* U of A s.t. $\overline{U} \cap B = \phi$.

Proof. The proof of necessity is as follows. Suppose that X is a normal space. Let A and B be disjoint non-empty closed subsets of X. Then $\exists$ open *nbds* V and W of A and B respectively, such that $V \cap W = \phi$. By virtue of Theorem 5.25, there exists an open set $U \supseteq A$ s.t. $\overline{U} \subseteq V$. Hence we have $\overline{U} \cap B = \phi$.

Now for sufficiency assume that the given condition holds. If we set $V = (\overline{U})'$, then U and V are disjoint open *nbds* of A and B respectively. Hence X is normal. $\square$

Theorem 5.28. For a T_1–space X to be normal it is necessary and sufficient that given any disjoint non-empty closed subsets A and B of X, $\exists$ open *nbds* U and V of A and B respectively such that $\overline{U} \cap \overline{V} = \phi$.

Proof. Necessity is shown by assuming that X is a normal space. Let A and B be disjoint non-empty closed subsets. Since B' is an open *nbd* of A, by virtue of Theorem 5.27, there exists an open set $U \supseteq A$ s.t. $\overline{U} \subseteq B'$. Applying the same theorem again to B and its open *nbd* $(\overline{U})'$ we get an open set $V \supseteq B$, s.t. $\overline{V} \subseteq (\overline{U})'$. Then U and V are open *nbds* of A and B respectively such that $\overline{U} \cap \overline{V} = \phi$. Sufficiency is trivial. This establishes our assertion. $\square$

Theorem 5.29. A T_1–space X is normal iff every closed set in X has a neighborhood bases of closed neighbourhoods.

Proof. The proof is left as a simple exercise. $\square$

Theorem 5.30. Every compact Hausdorff space is normal.

Proof. Let X be a compact Hausdorff space. Given disjoint closed sets A and B in X, choose for each point a of A, disjoint open sets U_a and V_a containing a and B respectively (see Chapter 6). The collection $\{U_a : a \in A\}$ is an open covering of A, and since A being a closed subset of a compact space is itself compact, $\exists$ finitely many sets $U_{a_1}, \cdots, U_{a_n}$ s.t. $A \subseteq U_{a_1} \cup U_{a_2} \cup \cdots \cup U_{a_n}$. Then $U = \bigcup_{i=1}^{n} U_{a_i}$ and $V = \bigcap_{i=1}^{n} V_{a_i}$ are disjoint open neighbourhoods of A and B respectively. Hence X is normal. Note that in general regular spaces need not be normal, but under certain conditions they are normal. $\square$

A topological space X is called *Lindelöf* if every open cover of X has a countable subcover.

Example 6. The set of real numbers $\mathbb{R}$ with discrete topology is not a Lindelöf space because the open cover $\{\{x\} : x \in \mathbb{R}\}$ has no countable subcover.

Example 7. Consider the set $\mathbb{R}$ with the usual topology. Since $\overline{Q} = \mathbb{R}$, i.e., $\mathbb{R}$ is separable, the collection $\{(x - \frac{1}{n}, x + \frac{1}{n}) : x \in Q, n \in \mathbb{N}\}$ is a countable bases for the usual topology on $\mathbb{R}$, and hence $\mathbb{R}$ is a Lindelöf space.

Theorem 5.31. Every second countable space X is Lindelöf.

Proof. Let $\mathfrak{F}$ be an open cover of X and let β be a countable base of X. Then for each $x \in X$, there is a set $U_x \in \mathfrak{F}$ and $B_x \in \beta$ s.t. $B_x \subseteq U_x$. Since $\{B_x : x \in X\}$ is actually a countable family and covers X, there is a corresponding countable collection of the U_x's that covers X. □

Theorem 5.32. Every regular space with a countable base is normal.

Proof. Suppose that A and B are disjoint closed subsets of X. By regularity of X, for each $a \in A, \exists$ an open set U_a containing a s.t. $U_a \cap B = \phi$ and for each $b \in B, \exists$ an open set V_b containing b s.t. $V_b \cap A = \phi$. Clearly, $A \subseteq \bigcup\{U_a : a \in A\}$ and $B \subseteq \bigcup\{V_b : b \in B\}$. Since X is a Lindelöf space by virtue of Theorem 5.31, both A and B are Lindelöf (because subspaces of Lindelöf spaces are Lindelöf). Therefore, we can find countable subcovers $\{U_n : n \in \mathbb{N}\}$ and $\{V_n : n \in \mathbb{N}\}$ for A and B respectively. The sets $U = \bigcup_{n \in \mathbb{N}} U_n$ and $V = \bigcup_{n \in \mathbb{N}} V_n$ are open sets containing A and B respectively, but they need not be disjoint. Therefore, we need to construct two open *nbds* of A and B respectively that are disjoint. For this, let $G_1 = U_1$ and $H_1 = V_1 \setminus \overline{G_1}$. Let $G_2 = U_2 \setminus \overline{H_1}$ and $H_2 = V_2 \setminus \overline{G_1 \cup G_2}$. In general, set $G_n = U_n \setminus \overline{(H_1 \cup H_2 \cup \cdots H_{n-1})}$ and $H_n = V_n \setminus \overline{(G_1 \cup G_2 \cup \cdots G_n)}$. Note that each G_n is open, being the difference of an open set U_n and a closed set. □

5.9 Exercises

1. Using the set of the letters of the alphabet, construct an *oligarchical* topology of a king and five courtiers and the masses consist of the other 20 letters of the alphabet.

2. Show that $\mathbb{R}$ with the right (or the left) ray topology is T_0 but not T_1.

3. Show that a closed subspace of a normal space is normal.

4. Prove that every locally compact Hausdorff space is regular.

5. A topological space X is said to be completely normal if every subspace of X is normal. Show that a space X is completely normal iff for every pair of subsets, A and B, with $\overline{A} \cap B = \phi = A \cap \overline{B}$, there are disjoint open sets U and V, containing A and B respectively.

6. Prove the following theorem (Urysohn's Lemma): A space X is normal iff for each pair A and B of disjoint closed subsets of X, there is a continuous function $f : X \to [0, 1]$ s.t.
$$f(a) = 0 \text{ for each } a \in A,$$
$$f(b) = 1 \text{ for each } b \in B.$$

7. Show that a regular, Lindelöf space X (i.e., every open cover of X has a countable subcover) is normal.

8. Let $X = \mathbb{R}$ and let the topology for X be determined by a base consisting of all half-open intervals of the form $[a, b)$, where $a < b$. Show that:

    (a) X is first countable but not second countable;

    (b) X is separable;

    (c) X is Lindelöf;

    (d) X is not metrizable.

9. Show that every completely normal space is normal but the converse need not hold.

10. Show that every subspace of a T_2–space is a T_2–space.

11. Suppose that X is an infinite T_2–space. Show that X has an infinite discrete subspace. (For instance, if $X = \mathbb{R}$ with the usual topology, then it has discrete subspaces $\mathbb{N}$ and $\mathbb{Z}$, while if X is equipped with the co-finite topology, then it has no infinite discrete subspace; hence T_2–condition is very essential.)

12. Let $X = \{a, b, c\}$ and let the topology $\tau = \{\phi, X, \{b\}, \{a, b\}, \{b, c\}\}$. Show that X is not a T_1–space. Is $\tau_1 = \{\phi, X, \{a, b\}, \{a\}, \{b\}\}$ also T_1? Justify!

13. Show that any finite T_1–space is a discrete space.

14. Show that the graph of a continuous function f of a topological space into a Hausdorff space Y, $G_r(f) = \{(x, f(x)) : x \in X\}$, is a closed subset of the product space $X \times Y$.

15. Let $X = \mathbb{N}$ and let τ be the odd-even topology on X generated by the sets $\{2n-1, 2n\}, n \in \mathbb{N}$. Show that:

    (a) X is second countable and hence first countable;

    (b) X is separable;

    (c) X is Lindelöf;

    (d) X is not a T_0–space.

16. Let $X = \mathbb{R}$ and let be τ the p–inclusion topology. Show that X is a T_0–space but not a T_1–space.

17. Let $X = \mathbb{R}^2$, with the topology generated by the base $\beta = \{(a,b) \times \mathbb{R} | a, b \in \mathbb{R}\}$. Show that X is not a T_1–space. Is X a T_0–space? Verify!

18. Prove that if X is a second countable normal space then it is metrizable.

19. Show that every metric space is completely normal.

20. For $\tau = \{\phi, X, U_n = (0, 1 - \frac{1}{n}), \text{ for } n = 2, 3, 4, \cdots \}$ on the open unit interval, show that:

    (a) X is not a T_0–space;

    (b) X is second countable and thus first countable;

    (c) X is connected;

    (d) X is locally connected.

21. Let $X = [-1, 1]$ and let the topology be generated from sets of the form $[-1, b)$ for $b > 0$ and $(a, 1]$ for $a < 0$. Then all sets of the form (a, b) are also open. Show that:

    (a) X is a T_0 but not a T_1–space.

    (b) X is compact [Hint: Since in any open covering, the two sets which include 1 and -1 will cover X.]

22. Show that T_2–property is a topological invariant.

Chapter 6

Compact Spaces

The original meaning of the word "compact" was a "covenant" or "agreement". This meaning led to an adjective signifying "closely knit" as opposed to "loosely scattered". Due to its use by American advertisers, in ordinary usage nowadays it is taken to indicate something small (in the sense of occupying little space) along with the connotation of containing everything required. This is, more or less, the significance of its *mathematical* usage. In Topology, however, "small" has no meaning by itself. Only in a metric space would it be possible to assign it a meaning. How can we extend the above, intuitive, concept to a general topological space (X, τ)? For that matter, why should we want to do so? To follow the answers to these questions it is necessary to see the origins of the concept in the development of calculus and its application in various other fields.

Calculus is based on considerations regarding sequences and series. Thus the basis of differential calculus is the procedure of taking limits of sequences while that of integral calculus is the procedure of taking limits of series. In either case we require that the sequence converges. Consider the simple harmonic sequence $\{\frac{1}{n} | n \in \mathbb{N} \smallsetminus \{0\}\}$. The limit, as $n \to \infty$, is clearly 0. Suppose we want to use such a sequence in a function with domain $(0, 1)$, for example. Obviously, we cannot use this limit as the function will not be defined at the limit point. Thus it is often relevant to know whether a sequence of elements of a set converge to a limit *within* the set. If we had taken $[0, 1]$ as the domain for our function, the convergence would have been guaranteed.

One of the commonest uses of calculus is to obtain the maxima and minima of a given function. However, if the domain of the function is open there is no guarantee that such values exist. In general, a continuous mapping from the reals to the reals will map closed intervals into closed intervals (regarding a singleton set as a closed interval of zero length). We can ask whether the range is bounded from above or not. If it is, the function attains its maximum value and otherwise it does not. A function, f, is said to be bounded from *above*, if $\exists\, u \geq f(x)\ \forall x \in D$, where D is the domain of f. The *supremum* of f is defined as the *least* upper bound

$$s \equiv \sup f(x) = \min_{u}\{u \geq f(x) | x \in D\}. \tag{6.1}$$

If D is a closed interval $\exists\, x^* \in D$ s.t. $f(x^*) = s$, i.e., the maximum of the function exists and is the supremum. Similarly, f is bounded from *below* if $\exists\, \ell \leq f(x)\ \forall x \in D$. The *infimum* of f is defined as the *greatest* lower bound

$$i \equiv \inf f(x) = \max_{\ell} \{\ell \leq f(x) | x \in \mathcal{D}\}. \tag{6.2}$$

Again, if D is a closed interval $\exists\, x_* \in D$ s.t. $f(x_*) = i$, i.e., the minimum of the function exists and is the infimum. Notice that we *could* have one of the two guaranteed to exist in a semi–closed interval but to have a *guarantee* for the existence of either we need the interval to be fully closed.

Now let us extend our discussion from $\mathbb{R}$ to $\mathbb{R}^n$. All sequences in $\mathbb{R}^n$ can be regarded as n independent sequences in $\mathbb{R}$. Hence, if D is closed in each $\mathbb{R}$, i.e., if D is the Cartesian product of n closed intervals, all sequences in D will converge to a limit point in D. Further, for functions of n variables, $\mathbf{x} \equiv x^i$ ($i = 1, \cdots, n$), we will be guaranteed the existence of $\mathbf{x}^*$ and $\mathbf{x}_*$ s.t. $f(\mathbf{x}^*) = s$ and $f(\mathbf{x}_*) = i$. Hence the maximum and minimum can be determined.

One may be tempted to take, as the next generalization, that the domain considered should be fully bounded. However, this generalizations cannot work consistently. For example, consider the closed disc in $\mathbb{R}^2$. This is bounded and *is* a reasonable generalization of what we mean by a "compact" set for $\mathbb{R}$. However, consider its boundary, the circle $\mathbb{S}^1$, as the space under discussion. It is an unbounded set in the sense that $\partial \mathbb{S}^1 = \phi$. However, it is obviously a space that should be regarded as "compact". Essentially, we need to be able to break our space up into a finite number of pieces, for each of which we can construct something like the closed interval. The same argument holds for a closed sphere in $\mathbb{R}^3$, which is "compact" and its boundary, $\mathbb{S}^2$, which is an unbounded "compact" space. It is this concept that we need to generalize to be able to use the space under consideration for calculus in general and optimization of functions in particular. As we shall shortly see, there are many different generalizations of the concept which will need to be studied separately and with reference to each other.

6.1 The Standard Definition of Compactness

There were various attempts to extend the intuitive concept just explained. All except one of these definitions is qualified by an adjective. That one, standard, definition comes from a property of closed, bounded subsets of $\mathbb{R}$. These are, after all, the sets we are trying to extend to arbitrary topological spaces. This property, is provided by the Heine-Borel theorem, which states that for any closed, bounded $X \subsetneq \mathbb{R}$ every open cover has a finite sub-cover $\{U_i\}$ ($i = 1, \cdots, n$) s.t. $U_i \in \tau_u\ \forall i$ and $\bigcup_{i=1}^{n} U_i = X$. To be able to give the definition of compactness which is standard we need, first, a couple of other definitions. A collection of sets $\mathcal{O} = \{U_i\}_{i \in I}$ is said to be an *open cover* (or a *covering*) of a topological space (X, τ) if $U_i \in \tau\ \forall i \in I$ and $\bigcup_{i=1} U_i = X$. A subset of $\mathcal{O}$ whose union also gives X is called a *sub-cover*.

Example 1. For $(\mathbb{R}, \tau_u)$ we can choose $\mathcal{O}_1 = \{(-n, n) | n \in \mathbb{N}\}$ or $\mathcal{O}_2 = \{(n - 1, n + 1) | n \in \mathbb{N}\}$. Then $\mathcal{O}_3 = \{(-2n, 2n) | n \in \mathbb{N}\}$ is a sub-cover of $\mathcal{O}_1$. See Figure 6.1.

$\mathcal{O}_1$: $(-1,1)$ $(-2,2)$ $(-3,3)$ $(-4,4)$ $(-5,5)$ $(-6,6)$ $(-7,7)$ $(-8,8)$ $(-9,9)$ ...

$\mathcal{O}_3$: $(-2,2)$ $(-4,4)$ $(-6,6)$ $(-8,8)$ ...

Figure 6.1: Some members of $\mathcal{O}_1$ and $\mathcal{O}_3$ are displayed. As expected, $\mathcal{O}_3 \subset \mathcal{O}_1$.

Example 2. For $(\mathbb{R}, \tau_d)$ we can again take $\mathcal{O}_1$ and $\mathcal{O}_3$. Alternatively, we can choose $\mathcal{O}_4 = \{\{x\}|x \in \mathbb{R}\}$.

Example 3. For $(\mathbb{R}, \tau_c)$ we can define $\mathcal{O}_c = \{\{x\}'|x \in \mathbb{R}\}$. As an exercise check whether $\{\{x,y\}'|x \in \mathbb{R}\}$ is a sub-cover of $\mathcal{O}_c$.

Example 4. For $\mathbb{N}$ with the topology $\tau = \{\phi, U_i | i \in \mathbb{N}\}$ where $U_i = \{i, i+1, \cdots, i+j, \cdots\}$, we can take $\mathcal{O} = \{U_i\}_{i \in \mathbb{N}}$. Clearly, if we leave out U_0 we will not have a covering of $\mathbb{N}$ as $\mathbb{N} \smallsetminus (\bigcup_{i=1}^{\infty} U_i) = \{0\} \neq \phi$. However, we could define the sub-cover $\{U_i | i = 2n, n \in \mathbb{N}\}$.

A topological space (X, τ) is said to be *compact* if every open *cover of* X contains a *finite* sub-cover. This is obviously a generalization of the Heine-Borel property of $\mathbb{R}$ (or $\mathbb{R}^n$). It is, in fact, the simplest generalization of the notion of "compactness" in $\mathbb{R}$ and yet proves adequate.

Example 5. Every finite topological space is compact since the entire topology is finite and hence an open cover, which is a subset of the topology, must be finite. Further, every space with a finite topology (even if the space itself is infinite) is a compact space. Thus every indiscrete space is compact. Even if the topology is not finite but it has a finite basis, it is compact.

Example 6. Every infinite discrete space is non-compact as the open cover $\mathcal{O} = \{\{x\}|x \in X\}$ has no finite sub-cover.

Example 7. $(\mathbb{R}, \tau_u)$ is non-compact since $\mathcal{O} = \{(-n, n)|n \in \mathbb{N}\}$ has no finite sub-cover. More generally, $(\mathbb{R}^n, \tau_u)$ is non-compact since $\mathcal{O} = \{S_m(\mathbf{0})|m \in \mathbb{N}\}$ has no finite sub-cover, where $S_m(\mathbf{0})$ is the open sphere of radius m and centre $\mathbf{0} = (0, \cdots, 0) \in \mathbb{R}^n$.

Example 8. (X, τ_c) is compact. To prove this consider $a_i \neq a_j \in X$. Then $\{a_i\}', \{a_j\}' \in \tau_c$ and $\{a_i\}' \cup \{a_j\}' = X$. Hence we have a 2–element open cover of X. Similarly, if we had $a_i, a_j, a_k, a_l \in X$ such that all elements are distinct we could use $\{a_i, a_j\}'$ and $\{a_k, a_l\}'$. This could be extended to any number of elements in 2 or more sets. Obviously, given any open cover we could select a finite subset from it which would also be an open cover. Hence every open cover has a finite sub-cover and the space is compact.

Example 9. The topological space (X, τ_{eA}) is compact, where τ_{eA} is the A–exclusion topology. This is so because every open cover in this space must contain X. If it did not, since $U_i \in \tau_{eA}$ is defined by the requirement $A \cap U_i = \phi$, $A \cap \left(\bigcup_{i \in I} U_i\right) = \phi$ and $\bigcup_{i \in I} U_i = X$ for an open cover, we would have $A \cap X = \phi$. Hence, for non-empty A we have a contradiction. Since every open cover must contain X, any finite subset containing X would be a finite sub-cover.

6.2 Characterizations of Compact Spaces

It may not always be easy to use the definition given above to verify if our space is compact or not. In such cases it is useful to provide alternative ways of testing compactness. As before we will present some theorems which give useful characterizations of the property of compactness.

Theorem 6.1. (X, τ) is compact iff every collection of closed sets, $\mathfrak{F} = \{F_\alpha\}_{\alpha \in I}$ s.t. $\bigcap_{\alpha \in I} F_\alpha = \phi$ has a finite sub-collection, $\mathfrak{G} = \{F_i\}_{i=1,\cdots,n} \subseteq \mathfrak{F}$ s.t. $\bigcap_{i=1}^n F_i = \phi$.

Proof. First assume that (X,τ) is compact and consider an $\mathfrak{F} = \{F_\alpha\}_{\alpha \in I}$ s.t. $\bigcap_{\alpha \in I} F_\alpha = \phi$. Define $U_\alpha = F'_\alpha$. By De Morgan's law $\bigcup_{\alpha \in I} U_\alpha = X$. Thus $\{U_\alpha\}_{\alpha \in I}$ is an open cover. Since (X,τ) is compact there exists a finite sub-cover $\{U_\alpha\}_{\alpha \in J}$ ($|J| = n \in \mathbb{N}$) s.t. $\bigcup_{i=1}^n U_i = X$. Again, by De Morgan's law $\bigcap_{i=1}^n U'_i = \phi$ for $\bigcap_{i=1}^n F_i = \phi$. Hence $\exists \; \mathfrak{G} = \{F_i\}_{i \in J} \subseteq \mathfrak{F}$ s.t. $\bigcap_{i=1}^n F_i = \phi$. This proves the necessity of the condition. To prove the sufficiency suppose that the condition holds. Since $\bigcap_{i=1}^n F_i = \phi$, we can use De Morgan's law to obtain $\bigcup_{i=1}^n F'_i = X$ or $\bigcup_{i=1}^n U_i = X$. Hence $\{U_i\}_{i \in J}$ is a finite sub-cover of X (where $\{U_\alpha\}_{\alpha \in I}$ is the cover). $\square$

A collection of subsets $\mathfrak{G} = \{A_\alpha\}_{a \in I}$ is said to possess the *finite intersection property* if for all finite sub-collections $\{A_i\}_{i \in J}$ ($|J| = n \in \mathbb{N}$), $\bigcap_{i \in J} A_i \neq \phi$. Here A_α need not be closed or open sets, but just subsets of X. This definition allows us to re-formulate the above characterization in a slightly different way which is convenient for actual use in many cases.

Theorem 6.2. (X,τ) is compact iff $\forall \mathfrak{G} = \{F_\alpha\}_{a \in I}$ s.t $F'_\alpha \in \tau$, which satisfy the finite intersection property, $\bigcap_{\alpha \in I} F_\alpha \neq \phi$.

Proof. This is obviously a re-statement of the previous theorem and so the previous proof, appropriately re-worded, applies. This re-wording is left as an exercise for you. $\square$

An example of a collection satisfying the finite intersection property is the nested sequence of closed sets $F_1 \supset F_2 \supset \cdots \supset F_i \supset F_{i+1} \supset \cdots$.

6.3 Construction of Compact Subspaces

It is not necessary that if a subspace is compact (non-compact) that the space will be compact (non-compact). In fact we can have non-compact subspaces of compact spaces as well as compact subspaces of non-compact spaces. Given a non-compact space we can look for the "smallest addition" to it to make it compact. The resultant space is called the *compactified space* and the addition is called a *compactification*. Here we go in the reverse direction. Given a space we want a compact subspace of it.

Theorem 6.3. $(Y,\tau_Y) \subseteq (X,\tau_X)$ is compact iff $\forall \mathcal{O} = \{U_\alpha\}_{\alpha \in I}$ s.t. $U_\alpha \in \tau_X$ and $\bigcup_{i \in J} U_i \supseteq Y$, $\exists \; \mathcal{O}_f \subseteq \mathcal{O}$ s.t. $\mathcal{O}_f = \{U_i\}_{i \in J}, |J| = n \in \mathbb{N}$.

Proof. Notice that $U_\alpha \in \tau_X$. Had we required that $U_\alpha \in \tau_Y$ we would have had a tautology. The reason why there is something to prove is that we require openness in the sense of the topology of the whole space and not of the subspace. Thus, for example, we could have $Y \notin \tau_X$. First suppose that Y is compact and $\mathcal{O}$ is some open cover of Y with its elements open in τ_X. Then $\mathcal{O}_Y = \{U_\alpha \cap Y\}_{\alpha \in I}$ is an open cover of Y in the sense of τ_Y. Since Y is compact, $\exists \; \mathcal{O}_{Yf} = \{U_i \cap Y\}_{i \in J} \subseteq \mathcal{O}_Y$ s.t. $|J| = n \in \mathbb{N}$. Hence $\exists \; \mathcal{O}_f = \{U_i\}_{i \in J}$ which is a finite sub-cover open in the sense of τ_X. Now suppose that the given condition holds and $\mathcal{O}_Y = \{V_\alpha\}_{\alpha \in I}$ is a cover of Y with $V_\alpha \in \tau_Y$. Hence $\forall \alpha \in I$ we can choose $U_\alpha \subseteq X$ s.t. $U_\alpha \cap Y = V_\alpha$. Now, since $\mathcal{O} = \{U_\alpha\}_{\alpha \in I}$ is a cover of Y, by our condition $\exists \; \mathcal{O}_f = \{U_i\}_{i \in J} \subseteq \mathcal{O}$ s.t. $|J| = n \in \mathbb{N}$. Hence $\exists \; \mathcal{O}_{Yf} = \{V_i\}_{i \in J}$. Therefore Y is compact. $\square$

Example 1. Obviously every finite subset of a topological space is a compact subspace, regardless of the topology chosen.

Example 2. $[a,b] \in \mathbb{R}$ is compact in the usual topology. To see this fact note that open sets, here, are simply open intervals. Every open interval has a finite length, howsoever small it may be. Any open cover of $[a,b]$ will include a finite number of intervals whose union is $[a,b]$, i.e., there will be a finite sub-cover of $[a,b]$. Remember that there *must* be some interval containing the point $\{a\}$ and some containing the point $\{b\}$. Had we, instead taken (a,b) this claim would not be valid. For example, for $a = 0, b = 1$, we could choose $\mathcal{O} = \{U_n\}_{n \in \mathbb{N}}$ given by $U_n = \left(\frac{1}{n+3}, \frac{n+2}{n+3}\right)$ for which there would be no finite sub-cover, see Figure 6.2.

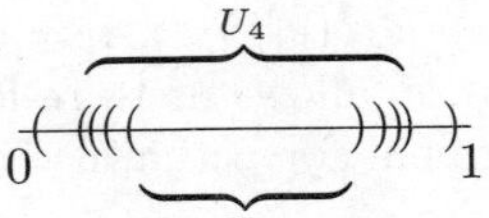

Figure 6.2: The open sets $U_n = \left(\frac{1}{n+3}, \frac{n+2}{n+3}\right)$ are depicted for $n = 1, 2, 3, 4$ and 20. For any arbitrarily large n, U_n will contain neither 0 nor 1. Hence a finite sub-cover for an open cover of $(0,1)$ cannot exist.

The above result, contained in Example 2, can be generalized to any arbitrary topological space *which is compact*. (In that example, of course, $\mathbb{R}$ is *non*-compact.) This fact is proved now.

Theorem 6.4. *Every closed subspace of a compact space is compact.*

Proof. For a compact (X, τ_X) consider $Y \subseteq X$ s.t. $Y' \in \tau_X$ and take $\mathfrak{F} = \{F_\alpha\}_{\alpha \in I}$ s.t. $F_\alpha \subseteq Y$, $F_\alpha \in \tau_Y$ (the induced topology on Y) s.t. $\mathfrak{F}$ satisfies the finite intersection requirement. As Y is closed in (X, τ_X), $\mathfrak{F}$ must also contain a collection of closed subsets, in the sense of τ_X, having the finite intersection property as well. Since X is compact, by Theorem 6.2, $\bigcap_{\alpha \in I} F_\alpha \neq \phi$. Hence Y is compact. □

Notice that we had to use the compactness of X. As such we have *not* proved that every closed subset is compact. It is worthwhile to try constructing an example of a closed subset that is not compact. Would it be a good idea to use the co-finite topology on $\mathbb{R}$ as the example? Can we generalize the result further? The answer is "yes, provided the space is Hausdorff", as seen in the next theorem.

Theorem 6.5. *Every compact subspace of a Hausdorff space is closed.*

Proof. Consider a Hausdorff space (X, τ_X) and a compact subspace $Y \subseteq X$. If $Y = \phi$ or X it is automatically closed (and open of course, but that is not relevant for us). As such we only need to consider a proper subspace. In this case $\exists\, x_0 \in X \smallsetminus Y$, see Figure 6.3. Since X is Hausdorff, $\forall y \in Y\ \exists\, U_y \ni x_0, V_y \ni y$ s.t. $U_y \cap V_y = \phi$ and $U_y, V_y \in \tau_X$. Obviously $\mathcal{O} = \{V_y | y \in Y\}$ is a covering of Y. (Provided we take care to choose $V_y \subseteq Y$, the covering would not extend outside Y, as otherwise $\bigcup_{y \in Y} V_y \supseteq Y$.) Thus $\exists\, \mathcal{O}_f \subseteq \mathcal{O}$ s.t. $|\mathcal{O}_f| = n \in \mathbb{N}$. Now let $\mathcal{O}_f = \{V_{y_1}, \cdots, V_{y_n}\}$. Then $V = \bigcup_{i=1}^n V_{y_i} \in \tau_X$. Also $U_{x_0} = \bigcap_{i=1}^n U_{y_i}$ is an open *nbd* of x_0 s.t. $V \cap U_{x_0} = \phi$. Hence $\forall x_0 \in X \smallsetminus Y$, we can construct open *nbds* U_{x_0} s.t. $U_{x_0} \cap V(\supseteq Y) = \phi$. Hence $U_{x_0} \cap Y = \phi$. Hence $x_0 \notin \overline{Y}\ \forall x_0 \notin Y$. Thus $\overline{Y} = Y$. □

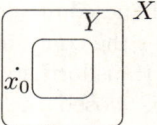

Figure 6.3: As $Y \subset X$, $x_0 \in X \smallsetminus Y$ will live in the annular region as shown.

With reference to the question raised above (in connection with the counterexample), it would be useful to incorporate the Hausdorff requirement as well. Remember that $(\mathbb{R}, \tau_c)$ is not Hausdorff. It is also worth noting that this theorem proves that (a,b) is non-compact [since $(\mathbb{R}, \tau_u)$ *is* Hausdorff] and $[a,b]$ is compact. Further, in a co-countable space the *only* compact subsets are finite since, if $A \subseteq X$ is infinite it is either discrete or co-countable. Neither of these are compact. Note that the co-countable space is also non-Hausdorff.

6.4 Theorems About Compact Spaces and Compactness

Compactness is a very stringent condition. As such it leads to various powerful statements that can be made about spaces possessing this property. Further, there are powerful statements to be made about mappings between compact spaces. In this section we state some theorems in this regard. Bear in mind that "compactness" arose as an attempt to extend certain properties of closed intervals to more general spaces. As we have seen, not all closed subspaces of topological spaces need be compact, nor must all compact subspaces be closed. As such "compactness" and "closedness" are very distinct topological properties though they both come from the closed interval in $(\mathbb{R}, \tau_u)$.

Theorem 6.6. Every compact Hausdorff space is normal and hence regular.

Proof. For a space (X, τ) to be normal we require that $\forall F, G \subseteq X$ s.t. $F', G' \in \tau$ and $F \cap G = \phi, \exists\, U \supseteq F, V \supseteq G$ s.t. $U, V \in \tau$ and $U \cap V = \phi$. The proof is essentially the same as for Theorem 6.5, with x_0 replaced by F. Now we require each U_y to contain F instead of only x_0, i.e., $F \subseteq U_y$. By definition, since the compact Hausdorff space is normal, it is regular. □

Theorem 6.7. Continuous mappings map compact spaces to compact spaces.

Proof. Let (X, τ_X) be a compact topological space and (Y, τ_Y) be the topological space it is mapped to. We can assume $f(X) = Y$. [If $f(X) = V \subsetneq Y$, we could consider (V, τ_V) as the space with τ_V being the topology inherited from τ_X.] Consider any $\mathcal{O}_Y = \{V_\alpha | \alpha \in I\}$, which is an open covering. Since f is continuous $U_\alpha = f^{-1}(V_\alpha) \in \tau_X$ and hence $\mathcal{O}_X = \{U_\alpha | \alpha \in I\}$. Since X is compact $\exists\, \mathcal{O}_{Xf} = \{U_i | i = 1, \cdots, n\}$ and hence $\mathcal{O}_{Yf} = \{V_i | i = 1, \cdots, n\}$. Thus Y is compact. □

This is a vitally important theorem as it automatically implies that compactness is a topological property. Without this fact we would not have been particularly interested in the study of compactness.

Theorem 6.8. A continuous bijection from a compact Hausdorff space is a closed mapping and hence a homeomorphism.

Proof. Consider any closed subset, A, of the compact space X. By Theorem 6.4 it is compact and hence, by Theorem 6.6, its image under the continuous bijection f, $B = f(A)$, is also compact. Thus, by Theorem 6.5, B is closed in the Hausdorff space Y. Hence f is a closed mapping. Since the pre-image of a closed set, $f^{-1}(B)$, is closed the inverse is also closed. Therefore f is a homeomorphism. □

Theorem 6.9. *The product of finitely many compact spaces is compact.*

Proof. The natural line of reasoning to take would be to consider any open cover of the product space, write it as products of open sets of the spaces whose product is being taken and use the compactness of those spaces. There is a problem with this simple attempt. Consider $(X, \tau) = (X_1, \tau_1) \times (X_2, \tau_2)$. It is not necessary that every $U \in \tau$ can be written as $U_1 \times U_2$ with $U_1 \in \tau_1$ and $U_2 \in \tau_2$. To see this fact concretely let $(X_1, \tau_1) = (X_2, \tau_2) = (\mathbb{R}, \tau_u)$ and $(X, \tau) = (\mathbb{R}^2, \tau_u)$. Consider the open cover of the product space by open discs $\mathcal{O} = \{U_{(n,m)} | U_{(n,m)} = \{(x,y) | (x-n)^2 + (y-m)^2 < 4, n, m \in \mathbb{Z}\}\}$. Now $\mathcal{O}_1 = \mathcal{O}_2 = \{U | U = (a,b), a < b \in \mathbb{R}\}$ will not allow us to write $U_{(n,m)} = U_1 \times U_2$. To avoid this problem, we could take any open cover of X_1 and X_2 respectively. Since these spaces are compact $\exists\ \mathcal{O}_{1f} \subseteq \mathcal{O}_1$ and $\mathcal{O}_{2f} \subseteq \mathcal{O}_2$, which are finite. Then we can define $\mathcal{O}_f = \{U_{\alpha_1} \times U_{\alpha_2} | U_{\alpha_1} \in \mathcal{O}_{1f}, U_{\alpha_2} \in \mathcal{O}_{2f}\}$. Since these are collections of finitely many open sets, the product contains finitely many open sets which provide a cover for the entire space. Hence X is compact. Proceeding pair-wise we could then multiply this product space by (X_3, τ_3) and then by (X_4, τ_4) and so on. Hence the product of finitely many compact spaces, $\prod_{i=1}^{n}(X_i, \tau_i)$, is compact. □

6.5 Compactness in Metric Spaces

So far our discussion has been entirely general. It becomes much clearer what compactness entails if we limit our discussion to metric spaces. We obtain many further characterizations of compactness and can deal with various special types of properties associated with compactness in such spaces. Of course, not all the properties we will be referring to would apply for non-metrizable spaces. Certainly the geometric (and hence analytic) significance of compactness can only emerge from metric spaces.

Our first characterization of compactness emerges from a generalization of the Bolzano-Weierstrass theorem: "Every infinite subset A, of a closed and bounded subset X, of $\mathbb{R}$ has a limit point in X". This theorem does not hold for all arbitrary closed and bounded sets X. Those X for which it does hold are said to possess the *Bolzano-Weierstrass property*. A metric space possessing this property is said to be *limit point compact*. [It is interesting to note that this was the original definition of "compactness", back in the early days of Topology when the concepts were being formulated. At that time the current definition in terms of open covers (which happens to come from generalizing the Heine-Borel theorem) was referred to as "bicompactness".] It should be stressed that the definitions coincide for metric spaces (as will be shown shortly) but there are spaces for which the two definitions are not equivalent.

Theorem 6.10. *Every compact metric space is limit point compact.*

Proof. We need to prove that every infinite subset A, of the compact metric space X, has a limit point in X. We do so by *reductio ad absurdum*. Suppose that the statement is false. For concreteness think of $A \subsetneq X$. As no $x_i \in X$ is a limit point of A, each must be the centre of an open set of arbitrary radius r_j, $S_{r_j}(x_i)$, which contains no other point of A, i.e., $S_{r_j}(x_i) \cap A = \{x_i\}$. Now $\{S_{r_j}(x_i) | x_i \in X, r_j > 0\} = \mathcal{O}$ is an open cover of X. Since X is compact $\exists\, \mathcal{O}_f \subseteq \mathcal{O}$ which is finite, $\mathcal{O}_f = \{S_{r_j}(x_i) | i = 1, \cdots, n\}$. Hence $A = \bigcup_{i=1}^n \{x_i\}$. So A must be finite, in contradiction to our original assumption that A is an *infinite* subset of X. $\square$

The second characterization arises from the requirement that every sequence in the metric space has a convergent subsequence. The metric space is then said to be *sequentially compact*. What is the relationship between this characterization and the earlier one for limit point compactness? For metric spaces the two are equivalent. This is demonstrated in the next theorem.

Theorem 6.11. A metric space is sequentially compact iff it is limit point compact.

Proof. (Necessity.) Suppose X is a sequentially compact metric space. Consider an infinite subset of it, A. Construct a sequence $\{x_n\}$ of distinct points of A. Hence $\exists$ a subsequence $\{x_{n_k}\}$ which converges to some point, say $x \in X$. Then, obviously x is a limit point of the set $B = \{x_{n_1}, \cdots, x_{n_k}, \cdots\}$, formed by the elements of the subsequence, i.e., $x \in \overline{B}$. Since $B \subseteq A \Rightarrow \overline{B} \subseteq \overline{A}$ we have $x \in \overline{A}$. Hence necessity is proved.

(Sufficiency.) Now consider the case that X is limit point compact. A subset $A = \{x_n | n \in \mathbb{N}\} \subseteq X$ is either finite or infinite. In the former case there exists some $x = x_n$ which is infinitely many times repeated. (There could be many such, but that makes no difference. We only need to consider one.) Hence A has a constant subsequence and therefore converges. If A is infinite it has a limit point (by assumption), x. We can now construct a sequence of open spheres: $\{S_1(x), \cdots, S_{1/k}(x), \cdots\}$, as shown in Figure 6.4. Since x is a limit point of A, let $x \in S_{1/k}(x)$. Hence we can choose

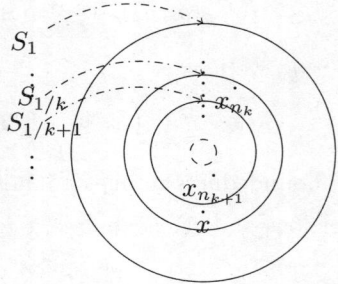

Figure 6.4: Construction of a sequence of open spheres is displayed. The limit point of the set $A, x \in S_{1/k}$. Hence the subsequence $\{x_{n_k}, x_{n_{k+1}}, \ldots\}$ contained in $S_{1/k}$ converges to x.

$x_{n_k} \in S_{1/k}(x)$ and $x_{n_{k+1}} \in S_{1/k+1}(x)$ with $n_{k+1} > n_k$. Hence the subsequence $\{x_{n_k}\}$ converges to x. This completes the proof. $\square$

We started our discussion of compactness by considering bounded and unbounded sets. Within the context of bounded sets can we have more or less bounded sets? One would have thought that either a set would be bounded or not. However, the Russian prodigy of the late 1960s, A.V. Arhangelskii (pronounced Arkhangelsky), provided a definition which did allow us to talk of more and less bounded sets. In a metric space, X, for given $\epsilon > 0$, he called a finite $A \subseteq X$ an ϵ–net if $\forall x \in X \; \exists \, a \in A$ s.t. $\rho(x,a) < \epsilon$. Thus, if $A = \{a_i | i = 1, \cdots, n\}$ and $X = \bigcup_{i=1}^{n} S_\epsilon(a_i)$, then A is an ϵ–net. See Figure 6.5. We say that X is *totally bounded* if it has an ϵ–net. This is a stronger

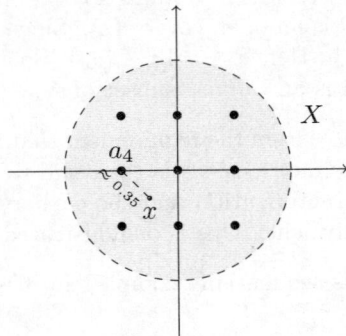

Figure 6.5: Let X be the open unit disk. If $\epsilon = 1/2$, the 9 black dots form an ϵ–net for X, given by $A = \{a_i | \rho(x, a_i) < \frac{1}{2} \; \forall x \in X\}$. However if we take $\epsilon = 1/4$, then A is not an ϵ–net for X. Consider an arbitrary member of A, say a_4, and some $x \in X$ that lies on the hypotenuse of a triangle formed by any three members of A. Then a simple calculation shows that $\rho(x, a_4) \not\leq 1/4$.

condition than boundedness as is apparent from the following argument. We call the supremum of all distances, $\rho(a,b)$, of points a,b of a set A, its *diameter* $d(A)$. (This will be defined only if the supremum exists.) Since our present A is finite, $d(A)$ is finite. Now $\forall x, y \in X$ we have

$$\begin{aligned}
\rho(x,y) &\leq \rho(x, a_i) + \rho(a_i, y) \\
&\leq \rho(x, a_i) + \rho(a_i, a_j) + \rho(a_j, y) \\
&\leq \epsilon + \rho(a_i, a_j) + \epsilon \leq d(A) + 2\epsilon.
\end{aligned} \qquad (6.3)$$

Hence, taking the supremum of $\rho(x,y)$ we have

$$d(X) \leq d(A) + 2\epsilon. \qquad (6.4)$$

Therefore X is bounded and total boundedness implies boundedness.

Example 1. $(\mathbb{R}, \tau_d)$ has no ϵ–net for $\epsilon < 1$. For example, take $\epsilon = \frac{1}{2}$. Now,

$$\mathbb{R} = \bigcup_{i=1}^{n} S_{1/2}(a_i). \qquad (6.5)$$

Since $S_{1/2}(a_i) = \{a_i\}$ in this topology and $\mathbb{R} \neq \{a_i\}$, this is impossible. [What would happen if we removed the restriction $\epsilon < 1$? As a hint, remember that A must be finite.]

Example 2. $(\mathbb{R}, \tau_u)$ has no ϵ–net, as for any $A = \{a_1, \cdots, a_n\}$ with $a_1 < \cdots < a_n$, $\bigcup_{i=1}^{n} S_\epsilon(a_i) \subseteq (a_1 - \epsilon, a_n + \epsilon) \neq \mathbb{R}$ for any finite ϵ.

Example 3. $((a,b), \tau_u)$ is totally bounded. To prove this we need to show that for *any* $\epsilon > 0$, $\exists$ an ϵ–net A. Consider some given ϵ. Now choose an integer $n \geq (b-a)/\epsilon$. Put $a_0 = a, a_1 =$

$a_0 + (b-a)/n$, $a_2 = a_1 + (b-a)/n$ and so on. Then $b = a_n$. Clearly, $A = \{a_1, \cdots, a_n\}$ is an ϵ-net. [What about $([a,b], \tau_u)$? The problem, in this case arises if $n = (b-a)/\epsilon$ in the above proof. Then $S_\epsilon(a_1) \supseteq a_0 = a$. Is this a serious problem?]

So far we have only proved that total boundedness implies boundedness. This is insufficient to prove our claim that the former is a *stronger* criterion than the latter. We need an example of a bounded set that is *not* totally bounded. For this purpose consider an infinite dimensional vector space

$$X = \left\{ \mathbf{x} = (x_1, \cdots, x_i, \cdots, x_n) | i \in \mathbb{N}; x_i \in \mathbb{R}; \mathbf{x} \cdot \mathbf{x} = \sum_{i=1}^{n} x_i^2 < \infty \right\}, \tag{6.6}$$

with the metric

$$\rho(\mathbf{x}, \mathbf{y}) = \sqrt{(\mathbf{y} - \mathbf{x}) \cdot (\mathbf{y} - \mathbf{x})}, \tag{6.7}$$

and the topology generated by this metric. Consider the set of unit basis vectors for this space

$$B = \{\mathbf{e}_n | n \in \mathbb{N}, e_{n_i} = 1 \text{ if } i = n \text{ and } e_{n_i} = 0 \text{ if } i \neq n\}, \tag{6.8}$$

where e_{n_i} is the ith component of the $\mathbf{e}_n$. Clearly we have

$$\rho(\mathbf{e}_m, \mathbf{e}_n) = \sqrt{2} \ \forall \ m, n \in \mathbb{N}, \tag{6.9}$$

and hence B is bounded, with diameter $\sqrt{2}$. However, B is *not* totally bounded because for any $\epsilon < \sqrt{2} \ \nexists$ an ϵ-net. The problem arises because of the fact that we would need an infinite set A to "catch" *all* the elements of B in its net, since B has all its elements placed along different "directions" and has infinitely many such. It is the fact of the infinitely many basis vectors that brings about the distinction. In fact, for a finite dimensional Euclidean space, $\mathbb{R}^n (n \geq 1)$, the two criteria coincide. Further, since every finite dimensional Riemannian manifold is locally homeomorphic to $\mathbb{R}^n$ (for some n), the two criteria coincide on all finite dimensional Riemannian manifolds.

We are now in a position to state the theorem for which the definition of total boundedness was given.

Theorem 6.12. Every sequentially compact space is totally bounded.

Proof. In a metric space X for a given $\epsilon > 0$, choose some $x_1 \in X$. If $S_\epsilon(x_1) = X$ then $\{x_1\}$ is an ϵ-net. If not, choose $x_2 \in [S_\epsilon(x_1)]'$. If $S_\epsilon(x_1) \cup S_\epsilon(x_2) = X$ then $\{x_1, x_2\}$ is an ϵ-net. If not, choose $x_3 \in [S_\epsilon(x_1) \cup S_\epsilon(x_2)]'$. Define

$$A_1 = \{x_1\}, A_2 = \{x_1, x_2\}, \cdots, A_n = \{x_1, \cdots, x_n\}. \tag{6.10}$$

If no A_n forms an ϵ-net for a finite n we can construct a set

$$A_\infty = \{x_1, \cdots, x_n, \cdots\}. \tag{6.11}$$

The sequence $\{x_1, \cdots, x_n, \cdots\}$ has no convergent subsequence as $\rho(x_i, x_j) \geq \epsilon$ (by construction) $\forall \ i, j \ (j \neq i)$. A sequentially compact X can never contain such an A_∞. Hence for every sequentially compact metric space, for some n there exists an ϵ-net A_n. Thus every sequentially compact space is totally bounded. $\square$

Notice that we did not guarantee an *efficient* procedure of constructing the ϵ–net in which we can have the minimal value for n. [It would be useful, as an exercise, to construct a minimal ϵ-net for $[0, 1]$ with $\epsilon = 0.1$.] Notice that an infinite dimensional vector space is not sequentially compact as the sequence $\mathbf{e}_n$ never converges there.

We know that compactness is a stronger criterion than boundedness, in that all compact sets are bounded but not every bounded set is compact. How about total boundedness, which is also stronger than boundedness? From the above theorem we already know that it cannot be stronger than compactness. But can it be as strong, i.e., equivalent? A simple counterexample shows that it *is* weaker. For example $X = (0, 1]$ with the usual metric on $\mathbb{R}$ is totally bounded but the sequence $\{1/n | n \in \mathbb{N}\}$ has no convergent subsequence.

6.6 Equivalence of Definitions of Compactness

We have already seen that for metric spaces sequential compactness is equivalent to limit point compactness. The question that remains to be answered is how they are related to compactness for metric spaces. Since limit point compactness is implicit in compactness, the latter must be the stronger criterion if the two are inequivalent. In the subsequent theorem we shall prove that all three criteria are equivalent. Before proceeding to that theorem we need to prove a lemma. The lemma to be proved is stated in terms of a definition we need to provide first.

A real number $\delta > 0$ is called a *Lebesgue* (pronounced Le-Beg) *number* for a given open cover $\mathcal{O} = \{U_\alpha | \alpha \in I\}$ of a metric space X, if each subset of X whose diameter is less than δ is contained fully in at least one U_α. It is by no means necessary that every metric space have a Lebesgue number. In fact only sequentially compact spaces can be guaranteed to have a Lebesgue number. This fact is proved in the following result known as the *Lebesgue number lemma*.

Theorem 6.13. *Every open cover of a sequentially compact metric space has a Lebesgue number.*

Proof. We prove the theorem by *reductio ad absurdum*. Consider an open cover $\mathcal{O} = \{U_\alpha\}_{\alpha \in I}$ of a sequentially compact metric space X and suppose there is no Lebesgue number. By definition, $\forall n \; \exists \; B_n \subseteq X$ s.t. $d(B_n) < 1/n$ and $B_n \not\subseteq U_\alpha$ for any $\alpha \in I$. Thus we can choose a sequence $\{x_n\}$ of distinct $x_n \in B_n$ (see Figure 6.6). Since X is sequentially compact $\exists \; \{x_{n_k}\} \subseteq \{x_n\}$ which converge to some $\overline{x} \in X$. Let $\overline{x} \in U_\alpha \in \mathcal{O}$ and define $F_\alpha = U'_\alpha$. Then $\epsilon = \rho(\overline{x}, F) > 0$. By the triangular inequality, $\forall x \in B_{n_k}$ we have

$$\rho(x, \overline{x}) \leq \rho(x, x_{n_k}) + \rho(x_{n_k}, \overline{x}) < 1/n_k + \rho(x_{n_k}, \overline{x}). \qquad (6.12)$$

By choosing a sufficiently large k we can ensure that

$$1/n_k < \epsilon/2 \text{ and } \rho(x_{n_k}, \overline{x}) < \epsilon/2. \qquad (6.13)$$

Therefore $\rho(x, \overline{x}) < \epsilon$. From the definition of ϵ and the openness of U_α, $S_\epsilon(\overline{x}) \subseteq U_\alpha$. Since U_α is open, $x \in U_\alpha \forall x \in B_{n_k}$. Thus

$$B_{n_k} \subseteq U_\alpha, \qquad (6.14)$$

contrary to our initial assumption. This completes the proof. $\qquad \square$

We are now in a position to prove the main claim of equivalence.

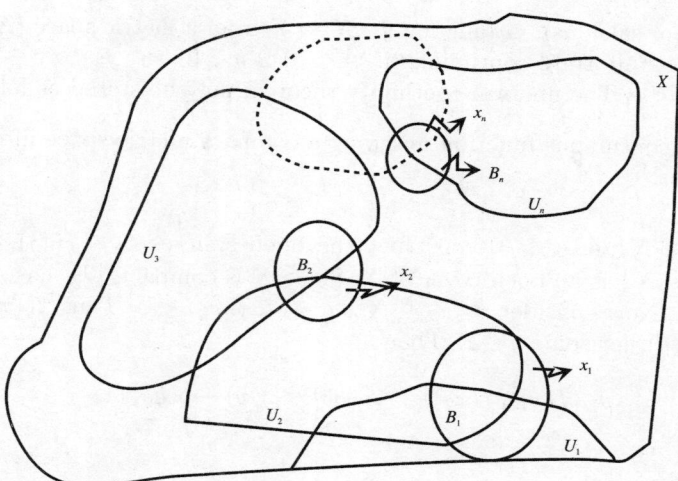

Figure 6.6: The space X has an open cover U_n. We construct a sequence of sets $\{B_n\}$ of decreasing diameters, $d(B_n) = 1/n$. If there is no Lebesgue number we can so construct them that no B_n lies completely in any U_α. Changing the open cover to some $\mathcal{O}_1 = V_\alpha$, we could always choose a new sequence of sets $\{C_n\}$, with the same properties. In each set we could select a distinct point. This is possible even if $B_m \subsetneq B_n$ for $m < n$, as we can take $x_n \in (B_n \smallsetminus B_m)$.

Theorem 6.14. For a metric space, X, the following are equivalent criteria:

(i) X is compact;

(ii) X is limit point compact;

(iii) X is sequentially compact.

Proof. In Theorem 6.10 we proved that (i)$\Rightarrow$(ii) and in Theorem 6.11 that (ii)$\Leftrightarrow$(iii). We will now demonstrate that (iii)$\Rightarrow$(i) and close the circle of implications to prove the equivalence of all three. Assuming (iii) any open cover $\mathcal{O}$ has a Lebesgue number δ (by Theorem 6.13). Since X is totally bounded (on account of being sequentially compact, by Theorem 6.12) we can choose an ϵ–net. Now each open sphere centred at x_k, $S_\epsilon(x_k)$ has diameter $d(S_\epsilon(x_k)) \leq 2\epsilon$ for all $k = 1, \cdots, n$. Taking $\epsilon < \delta/2$, the diameter is less than the Lebesgue number, δ. Hence we can choose $U_{\alpha_k} \in \mathcal{O}$ s.t. $S_\epsilon(x_k) \subseteq U_{\alpha_k}$. It follows that

$$X = \bigcup_{k=1}^{n} S_\epsilon(x_k) \subseteq \bigcup_{k=1}^{n} U_{\alpha_k}. \tag{6.15}$$

Hence there exists a finite sub-cover $\mathcal{O}_f = \{U_{\alpha_1}, \cdots, U_{\alpha_n}\}$ and hence X is compact. Since (i)$\Rightarrow$(ii)$\Leftrightarrow$(iii)$\Rightarrow$(i), all three are equivalent. □

The fact that we have different characterizations of compactness in metric spaces is very useful and enables us to prove many other important results. One application of it is the *theorem of uniform continuity* (of calculus). A function $f : X \to Y$ from a metric space (X, ρ_X) into another (Y, ρ_Y) is said to be uniformly continuous if $\forall \epsilon > 0 \; \exists \; \delta > 0$ s.t. $\forall x_1, x_2 \in X, \rho_X(x_1, x_2) < \delta \Rightarrow \rho_Y(f(x_1), f(x_2)) < \epsilon$. The uniform continuity theorem may be stated as follows.

Theorem 6.15. A continuous function mapping a compact metric space into a metric space is uniformly continuous.

Proof. For all $x \in X$ and $\epsilon > 0$ construct the open spheres $S_{\epsilon/2}(f(x)) \subseteq Y$. Then $\mathcal{O}_X = \{f^{-1}(S_{\epsilon/2}(f(x))|x \in X\}$ is an open cover of X. Since X is compact, $\mathcal{O}_X$ has a Lebesgue number (by Theorem 6.13). Now consider $x_1, x_2 \in X$ s.t. $\rho_X(x_1, x_2) < \delta$. Thus $d(\{x_1, x_2\}) < \delta$ and so $f(x_1), f(x_2) \in S_{\epsilon/2}(y)$, for some $y \in Y$. Then

$$\begin{aligned} \rho_Y(f(x_1), f(x_2)) &\leq \rho(f(x_1), y) + \rho(y, f(x_2)) \\ &< \epsilon/2 + \epsilon/2 = \epsilon. \end{aligned} \tag{6.16}$$

Hence f is uniformly continuous as claimed. $\square$

Of particular interest is the special case when $(Y, \rho_Y) = (\mathbb{R}, \rho_u)$. Thus every continuous real valued function defined on a compact metric space is uniformly continuous.

Another useful application of the earlier theorem is the *maximum value theorem* (of calculus), the *generalized Weierstrass theorem*.

Theorem 6.16. A real valued continuous mapping from a compact subspace of a metric space is bounded and attains its supremum and infimum.

Proof. Let $A \subseteq X$ be the compact subspace and f be the mapping. Denote the infimum and supremum, as before by i and s. The claim is that $\exists \; a, b \in A$ s.t. $f(a) = i$ and $f(b) = s$, so that $f(a) \leq f(x) \leq f(b)$. Now, since f is continuous and A is compact, by Theorem 6.7, $B = f(A)$ is a compact subspace of $\mathbb{R}$. Therefore, by Theorem 6.5, it is closed. Further, by the Heine-Borel theorem, it is bounded. If B has no maximum, $\mathcal{O}_B = \{(-\infty, y_i)|y \in B\} \cap B$ is an open cover of B. Since B is compact $\exists$ a finite sub-cover of B, $\mathcal{O}_{Bf} = \{(-\infty, y_1), \cdots, (-\infty, y_n)\} \cap B$. If $y_k = \max\{y_1, \cdots, y_n\}$ then it belongs to none of the sets in $\mathcal{O}_B$, which is absurd as they cover B. In the same way we get a contradiction by assuming that there is no minimum. Hence i and $s \in B$. Thus $\exists \; a, b \in A$ s.t. $a = f^{-1}(i)$ and $b = f^{-1}(s)$. This completes the required proof. $\square$

6.7 Compactness and Completeness

Since compactness is equivalent to sequential compactness, every Cauchy sequence in a compact metric space has a convergent subsequence. Thus compact metric spaces are necessarily complete. However, complete spaces need not be compact. For example, even $(\mathbb{R}, \tau_u)$ with the corresponding metric, ρ_u, generated by τ_u is complete but is not compact. What is missing in completeness is boundedness. *That* is the essence of compactness (in some sense). More precisely, as we see in the subsequent theorem, it is *total boundedness*.

Theorem 6.17. A metric space is compact iff it is complete and totally bounded.

Proof. We first prove necessity. By Theorem 6.12 a compact metric space is totally bounded and by the above argument is complete. Hence the necessity is obvious. Sufficiency is a trickier matter. It will clearly be easiest to prove sequential compactness from completeness (with total boundedness) and then appeal to Theorem 6.14. Let $\{x_n\}$ be a sequence of points of a complete and totally bounded metric space, X. Since X is totally bounded it has an ϵ–net and can, therefore be covered by finitely many unit open spheres. If $\{x_n\}$ is an infinite sequence then at least one of these open spheres, call it S_1, contains infinitely many points in $\{x_n\}$. Similarly, we can cover X by open spheres of unit diameter, i.e., radius $1/2$. Again, at least one of these open spheres, call it S_2, contains infinitely many points. We can continue these constructions for any arbitrary k to obtain an S_k of radius $1/k$. Define the index

$$N_{k+1} = \{n \in N_k | x_n \in S_{k+1}\}, \qquad (6.17)$$

where

$$N_1 = \{n \in \mathbb{N} | x_n \in S_1\}. \qquad (6.18)$$

Hence $N_1 \supset N_2 \supset \cdots \supset N_k \supset \cdots$. Thus given any n_k we can choose $n_{k+1} \in N_{k+1}$ s.t. $n_{k+1} > n_k$. Further, if $i, j \geq k$ then $x_{n_i}, x_{n_j} \in S_k$. By the nature of this construction it is a Cauchy sequence and hence has a convergent subsequence, $\{x_{n_k}\} \subseteq \{x_n\}$. Since X is complete $\{x_{n_k}\}$ must converge in it. Thus X is sequentially compact and therefore compact. $\square$

This theorem, due to none other than Hausdorff, has a useful corollary which we proceed to state as a theorem.

Theorem 6.18. A closed subset of a complete metric space is compact iff it is totally bounded.

Proof. The sufficiency is obvious from the previous theorem. The only point to be proved is that mere boundedness is inadequate. This is most easily provided by an example. The space defined by Eq. (6.6) is complete and has the subset defined by Eq. (6.8), in this chapter. The subset is non-compact (as it spreads out into infinitely many directions. However, it is bounded. It is the difference between boundedness and total boundedness on account of which compactness is lost. To see that there exists an open cover of B which has no finite sub-cover, take

$$\mathcal{O}_B = \{S_2(\mathbf{e}_m) | \mathbf{e}_m \in B\}. \qquad (6.19)$$

Hence boundedness is inadequate while total boundedness is adequate for ensuring compactness. $\square$

The above result is important to see why we insisted on *total* boundedness instead of mere boundedness. It also has as a direct consequence, the *Heine-Borel theorem*.

Theorem 6.19 (Heine-Borel Theorem). Every closed, bounded subset of $\mathbb{R}^n$ is compact.

Proof. Since absolute boundedness is equivalent to boundedness in $\mathbb{R}^n$, the result follows automatically from Theorem 6.18. $\square$

In fact the above theorem applies also to any finite dimensional Riemannian manifold. In particular $[a, b]$ and $\mathbb{S}^1 = \{(x, y) | x^2 + y^2 = 1\}$ are compact spaces. However, it is not true for pseudo-Riemannian manifolds.

While the concept of compactness is very useful it is frustrating that the very simplest space, $(\mathbb{R}, \tau_u)$, is non-compact. One needs a weaker criterion which brings this space "into the fold". This is provided in the next (and last) section of this chapter.

6.8 Local Compactness

The required weaker definition for compactness was provided by the famous Russian topologist, P. S. Alexandrov. He called a topological space, X, *locally compact at a point* $x \in X$ if $\exists$ a nbd U of x s.t. $\overline{U}$ is compact. If X is locally compact at all of its points, it is called *locally compact*. Clearly, every compact space is locally compact. That this *is* the required definition is clear from the consideration of $(\mathbb{R}, \rho_u)$. Since $\forall x \in \mathbb{R}$ we can choose $U = S_\epsilon(x), U = [x - \epsilon, x + \epsilon]$ is compact (as seen in Section 6.7) and so $\mathbb{R}$ is locally compact. Similarly, $(\mathbb{R}^n, \rho_u)$ is locally compact. It would be a good idea to check whether this statement holds true for an infinite dimensional space, such as was considered earlier. Again, consider $(\mathbb{R}, \tau_d)$. Here, $\forall x \in \mathbb{R}, \{x\} = \overline{\{x\}}$ is a compact *nbd* of x. Hence $(\mathbb{R}, \tau_d)$ is also locally compact. Since $(\mathbb{R}, \tau_c)$ is compact it, too, is locally compact.

Theorem 6.20. *Any closed subspace of a locally compact space is locally compact.*

Proof. The local compactness of the subspace will, of course, be with respect to the relative topology it inherits. Let X denote the space and Y the subspace. Consider any $x_0 \in Y$ and $U_0 \subseteq X$ s.t. $x_0 \in U_0$ and $\overline{U}_0$ is compact. Then $Y \cap \overline{U}_0$ is closed and compact in the relative topology. Since $V_0 = Y \cap U_0$ is open in the relative topology we only need to prove that $\overline{V}_0$ (the closure according to τ_Y) is compact. Now

$$\overline{V}_0 = \overline{Y \cap U_0} \subseteq \overline{Y} \cap \overline{U}_0 = Y \cap \overline{U}_0. \tag{6.20}$$

Since $Y \cap \overline{U}_0$ is closed and compact so is $\overline{V}_0$. This completes the proof. $\square$

We see that whereas compactness is not an inherited property, to some extent local compactness is. This fact is what makes this particular weakening of the concept of compactness so useful. In fact the restriction to closed subsets (whereby $\overline{Y}$ was equal to Y in the above equation) is not absolutely necessary. By imposing some constraints on the original space, X, we can even extend the above result to open subspaces. (After all, the open interval in $(\mathbb{R}, \tau_u)$ *is* locally compact.) This fact is demonstrated in the subsequent theorem.

Theorem 6.21. *Every open subset, A, of a compact Hausdorff space, X, is locally compact.*

Proof. Since a compact Hausdorff space is regular $\forall x_0 \in A \; \exists \; U_0$ s.t. $x_0 \in U_0 \in \tau$ and $\overline{U}_0 \subseteq A$. By the above theorem since $\overline{U}_0$ is compact, A is locally compact. $\square$

For a mapping from a compact space to yield a compact image it is enough to require that the mapping be continuous. However, this is not, by itself, enough to ensure that locally compact spaces be mapped to locally compact spaces. What is additionally required becomes clear in the next theorem.

Theorem 6.22. *If $f : X \to Y$ is continuous and open and X is locally compact while Y is Hausdorff, then $f(X)$ is locally compact.*

Proof. Consider some $y_0 \in Y$ and $x_0 \in f^{-1}(y_0)$. Since X is locally compact $\exists\, U_0 \subseteq X$ s.t. $x_0 \in U_0 \in \tau_X$ s.t. $\overline{U}_0$ is compact. Since f is open, $V_0 = f(U_0)$ is an open *nbd* of y_0. Further $f(\overline{U}_0)$ is compact and $V_0 \subseteq f(\overline{U}_0)$. Since $f(\overline{U}_0)$ is compact and Y is Hausdorff, $f(\overline{U}_0)$ is closed. Thus $\overline{V}_0 \subseteq f(\overline{U}_0)$ and hence V_0 is compact. Therefore, $\forall y_0 \in f(X) = Y$, $\exists\, V_0$ s.t. $y_0 \in V_0 \in \tau_Y$ and $\overline{V}_0$ is compact and so Y is locally compact. $\square$

The above theorem naturally leads on to an important result.

Theorem 6.23. Local compactness is a topological property.

Proof. If X is a locally compact space and $X \cong Y$, consider the homeomorphism $f : X \to Y$. Let $y_0 \in f(X) = Y$ and $x_0 \in f^{-1}(y_0)$. Since X is locally compact, $\exists\, U_0$ s.t. $x_0 \in U_0 \in \tau_X$ and $\overline{U}_0$ is compact. Since f is open $V_0 = f(U_0)$ is an open *nbd* of y_0. Further, $f(\overline{U}_0) \subseteq Y$ is compact. It is also closed as f is a closed mapping. In fact $V_0 \subseteq f(\overline{U}_0)$. Thus Y is locally compact. This completes the proof. $\square$

We finally state a theorem for product topologies to be locally compact.

Theorem 6.24. The product of finitely many locally compact spaces is itself locally compact.

Proof. Consider any $\mathbf{x} = (x_1, \cdots, x_i, \cdots, x_n) \in X = \prod_{i=1}^{n} X_i$, where each X_i is locally compact. Thus $\forall x_i \in X_i, \exists\, U_i$ s.t. $x_i \in U_i \in \tau_i$ and $\overline{U}_i$ is compact. Therefore $U = \prod_{i=1}^{n} U_i$ is a *nbd* of x s.t. $\overline{U} = \prod_{i=1}^{n} \overline{U}_i$ is compact. Hence X is locally compact. $\square$

As a direct consequence of this theorem we see that $\mathbb{R}^n$ and $\mathbb{C}^n$ are locally compact, as are $\prod_{i=1}^{n}(a_i, b_i), \prod_{i=1}^{n}[a_i, b_i), \prod_{i=1}^{n}(a_i, b_i], \prod_{i=1}^{n}[a_i, b_i]$, etc., where $a_i, b_i \in \mathbb{C}$ or $\mathbb{R}$. Further all Riemannian manifolds, real or complex, are locally compact provided they are finite dimensional. This is why local compactness is such a useful property.

6.9 Exercises

1. Show that a sequentially compact space is complete and totally bounded. Is total boundedness a topological property? Justify your answer.

2. Prove that a subspace of $\mathbb{R}^n$ is bounded iff it is totally bounded.

3. Show that a subspace A of a metric space X is totally bounded iff $\bar{A}$ is so.

4. Show that the continuous image of a compact metric space is compact.

5. Let X be a compact space. Show that any continuous real or complex valued function on X is bounded.

6. Prove that the set of real numbers $\mathbb{R}$, with a discrete metric is a complete space but not totally bounded, and hence it is not a compact space.

7. Consider $\mathbb{R}$ with the finite complement (i.e., co-finite) topology. Then $\mathbb{R}$ is a compact space. Show that a compact subspace of $\mathbb{R}$ need not be closed. [Hint: consider the set of integers, $\mathbb{Z}$, with the relative topology. Then $\mathbb{Z}$ is a compact subspace of $\mathbb{R}$ which is neither open nor closed.]

8. Let X be a a compact space. Show that each compact subspace A is closed in X iff each continuous bijection f from a compact space Y onto X is a homeomorphism.

9. Recall that a topological space is said to be first countable if it has a countable open base at each of its points. Prove that any first countable space X is a T_2–space if and only if every compact subspace A of X is closed.

10. Consider the Fort topology on $\mathbb{R}$, i.e., $\tau = \{A \subseteq \mathbb{R} | A' \text{ is finite or } p \notin A\}$, where p is some particular point of $\mathbb{R}$. Show that this space is compact.

11. Prove that $(0,1]$ is not compact. [Hint: The open covering $F = \{(\frac{1}{n}, 1] | n \in \mathbb{N}\}$ contains no finite sub-cover of $(0,1]$.]

12. Consider the subspace $X = \{0\} \cup \{1/n | n \in \mathbb{N}\}$ of $\mathbb{R}$ equipped with the relative topology inherited from the usual topology of $\mathbb{R}$. Show that X is compact.

13. Consider the usual topology on $\mathbb{R}$. Define a new topology $\sigma = \{A \subseteq \mathbb{R} | A = \phi \text{ or } A' \text{ is compact in the usual space } \mathbb{R}\}$. Note that since compact sets in $\mathbb{R}$ are closed under arbitrary intersection and finite union, so σ is in fact a topology on $\mathbb{R}$. Prove that $(\mathbb{R}, \sigma)$ is a compact space. [Hint: if $\{G_\alpha\}$ is an open covering of $\mathbb{R}$, then G'_α is compact in the usual space $\mathbb{R}$. Moreover, since each G_α is also open in the usual space $\mathbb{R}$, a finite number of elements of $\{G_\alpha\}$, say $G_{\alpha_1}, \cdots, G_{\alpha_n}$ must cover G'_{α_0}. Hence $G_{\alpha_0}, G_{\alpha_1}, \cdots, G_{\alpha_n}$ form a finite sub-cover of $\mathbb{R}$, and therefore $(\mathbb{R}, \sigma)$ is compact.]

14. Determine whether $\mathbb{R}$ with respect to $d(x,y) = \min\{1, |x-y|\}$ is compact or not?

Chapter 7

Connectedness

'Connectedness' is a fundamental topological property, like compactness which occupies a central position in both Topology and Analysis. It can be viewed as a generalization of the basic property of an interval: it is all in one piece. Likewise, a circle in the plane is all in one piece. In the study of calculus, there is a basic theorem about continuous function known as the 'Intermediate Value Theorem', which asserts that for a real-valued continuous function, f, defined on the interval $[a,b]$ $(a,b \in \mathbb{R})$, if t lies between $f(a)$ and $f(b)$, then $\exists\, c \in [a,b]$ s.t. $f(c) = t$. The basic property of $[a,b]$ on which this theorem depends, besides the continuity of f, is the property called connectedness.

The definition of a connected space was given by C. Jordan. However, a systematic study of connected spaces was initiated and developed by F. Hausdorff, B. Knastern, K. Kuratowski and P. Urysohn in the early years of the 20th century.

When the concept of connectedness is extended to arbitrary topological or metric spaces it turns out that there is more than one generalization possible. In this chapter we will study some elementary properties of some of these generalizations, namely: locally connected spaces. We will also study components of spaces and totally disconnected spaces.

7.1 Connected Spaces

To be able to define connected spaces we first define 'disconnected spaces'. A pair (A, B) of non-empty open subsets of a topological space X is called a *separation* of X iff

$$A \cup B = X, \quad A \cap B = \phi. \tag{7.1}$$

Equivalently, if A and B are disjoint, non-empty closed subsets of X s.t. $X = A \cup B$, then A and B effect a separation of X, since their complements would be open disjoint sets whose union is the whole space X. Thus if X is separated by subsets A and B, then A and B are both open and closed. If a subset of X is both open and closed we call it *clopen*. A space X is said to be *disconnected* if there exists a separation of it.

We are now in a position to define a connected space. A connected space is one which is not disconnected i.e., a topological space X is said to be connected if there does not exist a separation of X.

It follows immediately from the definition that if X is a connected space and if A, B are any two non-empty open (respectively closed) subsets of X s.t. $A \cup B = X$, then $A \cap B \neq \phi$.

There is another equivalent formulation of connectedness, given below, which is frequently useful.

Theorem 7.1. A topological space X is connected iff the only subsets of X which are both open and closed are the empty set and X itself.

Proof. We first prove the sufficiency. Assume that X is connected. Let A be any non-empty proper clopen subset of X, then $X = A \cup A'$. Therefore, the pair (A, A') is a separation of X which is not possible by the connectedness of X. Thus we conclude that either $A = \phi$ or $A = X$.

Conversely, if (A, B) is any separation of X, i.e., Eq. (7.1) is satisfied, then by definition $A \neq \phi, X$. Hence A is both open and closed. This proves the theorem. $\square$

We now present several examples of connected spaces.

Example 1. Let X be any non-empty set endowed with the indiscrete topology. Then obviously X is connected.

Example 2. Let X be any non-empty set consisting of two or more points with the discrete topology. Then clearly X is disconnected.

Example 3. Let $X = \{1, 2, 3\}$ with the topologies:

$$\begin{aligned}
\tau_1 &= \{\phi, \{1, 2\}, X\}, \\
\tau_2 &= \{\phi, \{1\}, X\}, \\
\tau_3 &= \{\phi, \{2\}, \{2, 3\}, X\}, \\
\tau_4 &= \{\phi, \{1\}, \{2\}, \{1, 2\}, X\}, \\
\tau_5 &= \{\phi, \{1, 2\}, \{1, 3\}, \{1\}, X\}.
\end{aligned} \qquad (7.2)$$

Then (X, τ_i) is a connected space for each $i = 1, 2, 3, 4, 5$.

Example 4. Let X be an infinite set with the co-finite topology. (For definiteness think of X as $\mathbb{R}$.) Then X is a connected space. For if there does exist a separation of X, namely, $A \cup B = X$ and $A \cap B = \phi$, then A and B are both open and closed in X. Hence they are both finite (by definition) and so is their union. This is not possible since X is infinite.

Example 5. $(\mathbb{R}, \tau_u)$ is connected. Similarly, any interval (a, b) in $\mathbb{R}$ with the relative topology inherited from τ_u on $\mathbb{R}$ is connected. The proof of this fact will follow from later results (see the Corollary 7.14.1).

Example 6. $\mathbb{R}^n (n \geq 1)$ endowed with the Euclidean topology (i.e., the topology induced by the usual metric) is a connected space (see the Corollary 7.14.1).

Example 7. Any open or closed ball or sphere in $\mathbb{R}^n$ is connected. In particular, the circle $\mathbb{S}^1 = \{(x, y) | x^2 + y^2 = 1\}$ is connected. Similarly, the torus $T^2 = \mathbb{S}^1 \times \mathbb{S}^1$ and the cylinder $\mathbb{S}^1 \times [0, 1]$ are connected spaces. (Again the proofs will follow from later results.)

Example 8. Let X be any non-empty set equipped with the p–inclusion topology. Then X is connected. In fact, if (A, B) is a separation of X, then both A and B are open in X and hence $p \in A \cap B$. This contradicts our assumption. Hence, X is connected.

Example 9. Define a topology τ on $\mathbb{R}$ as follows:

$$\tau = \{A \subseteq \mathbb{R} | A' \text{ is finite or } A' \ni p\}. \tag{7.3}$$

That is, τ is the minimal topology generated by the p-exclusion topology together with the co-finite topology. Every point x of $\mathbb{R}$, except p, is both open and closed. Therefore, $\{x\}$ and $\mathbb{R} \smallsetminus \{x\}$ separates $\mathbb{R}$. Hence

$$\mathbb{R} = \{x\} \cup (\mathbb{R} \smallsetminus \{x\}), \tag{7.4}$$

and thus $\mathbb{R}$ is disconnected.

Exercise 7.1. Determine whether a non-empty set with the p–exclusion topology is connected or not.

Exercise 7.2. Consider the set $\mathbb{Q}$ of rational numbers with the relative topology inherited from the usual topology on $\mathbb{R}$. Determine whether $\mathbb{Q}$ is connected or not.

Exercise 7.3. Consider $(\mathbb{R}, \tau_u)$. Define another topology τ^* on $\mathbb{R}$ by

$$\tau^* = \{A \subseteq \mathbb{R} | A = \phi \text{ or } A' \text{ is compact in } (\mathbb{R}, \tau_u)\}. \tag{7.5}$$

Determine whether $(\mathbb{R}, \tau^*)$ is connected or not.

Exercise 7.4. Let $X = \{x \in \mathbb{N} | x \geq 2\} = \{2, 3, 4, \cdots\}$ and let τ be the topology on X generated by sets of the form

$$U_n = \{x \in \mathbb{N} | x \text{ divides } n \text{ for } n \geq 2\} \tag{7.6}$$

That is,

$$\begin{aligned} U_2 &= \{2\}, U_3 = \{3\}, U_4 = \{2, 4\} \\ U_5 &= \{5\}, U_6 = \{2, 3, 6\}, U_7 = \{7\}, \\ U_8 &= \{2, 4, 8\} \end{aligned}$$

and so on. Prove that (X, τ) is connected.

7.2 Further Theorems for Connected Spaces

The following theorem gives a useful characterization of connectedness.

Theorem 7.2. A topological space X is connected iff there is no continuous function $f : X \to Y$ which is onto, where Y is the discrete space consisting of just two points $\{0, 1\}$.

Proof. We first prove sufficiency. Assume that X is connected. If $f : X \to Y$ is a continuous function of X onto Y, then the sets $f^{-1}(0)$ and $f^{-1}(1)$ are open in X and form a separation of X. Hence X is disconnected, which contradicts our assumption.

Conversely, assume that X is disconnected. Then $\exists$ a separation (U,V) of X. Let $f : X \to Y$ be a function defined by,

$$f(x) = \left\{ \begin{array}{ll} 0 & \text{if } x \in U \\ 1 & \text{if } x \in V \end{array} \right\}. \tag{7.7}$$

Then f is clearly continuous and maps X onto Y. $\square$

Corollary 7.2.1. A topological space X is connected iff every continuous function $f : X \to Y$ is a constant function, where $Y = \{0,1\}$ with the discrete topology.

Proof. To prove sufficiency assume that X is a connected space and f is a continuous function from X into Y. For any $x \in X$, the set $f^{-1}(f(x))$ is both open and closed, but cannot be empty. Hence $f^{-1}(f(x)) = X$ by Theorem 7.1. Thus $f(X) = f(x)$, i.e., f is a constant function.

Conversely, if A is any open and closed subset of X, then the characteristic function $\chi_A : X \to \{0,1\}$ given by

$$\chi_A(x) = \left\{ \begin{array}{ll} 1 & \text{if } x \in A, \\ 0 & \text{otherwise,} \end{array} \right\} \tag{7.8}$$

is continuous, with the co-domain carrying the discrete topology. Hence, by assumption χ_A is constant. This implies that either A is empty or equals X. Hence X is connected. This completes the proof. $\square$

Like compactness, connectedness is preserved by continuous functions.

Theorem 7.3. If f is a continuous function from a connected space X into a topological space Y, then $f(X)$ is connected.

Proof. Let X be connected, and for the sake of simplicity assume that $f : X \to Y$ is a continuous surjection, i.e., $f(X) = Y$. We show that Y is also connected. Let B be any open and closed subset of Y. By Theorem 7.1, we have to prove that either B is empty or equal to Y. Since f is continuous, $f^{-1}(B)$ is both open and closed in X. Since X is connected, so either $f^{-1}(B)$ is empty or equal to X. Hence $f(f^{-1}(B)) = B$ is either empty or equal to Y. Thus Y is connected. $\square$

An alternate proof is also available and would be useful to see. Assume that Y is disconnected, then by Theorem 7.2 there is a continuous function g from Y onto Z, where $Z = \{0,1\}$ with discrete topology. Hence the composite function, $g \circ f$, is continuous from X onto Z which contradicts the fact that X is connected.

Corollary 7.3.1. Connectedness is a topological property.

We now present some applications of this theorem.

Application 7.1. It is well-known that $(0,1)$ is homeomorphic to $\mathbb{R}$. As is shown below, $(0,1)$ is connected, therefore $(\mathbb{R}, \tau_u)$ is connected.

Application 7.2. Similarly, if $X = \mathbb{R}^2$, then any straight line L passing through the origin is a connected subspace of $\mathbb{R}^2$ by the same argument (see Figure 7.1), as is any other line in $\mathbb{R}^2$.

As an immediate consequence of Theorem 7.3, we have the following classical result familiar from elementary calculus. This result, known as the 'Intermediate Value Theorem', proves to be a key tool in establishing certain basic theorems in calculus.

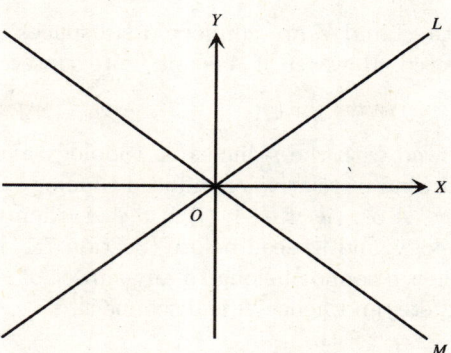

Figure 7.1: Lines L and M passing through the origin are connected subspaces of $\mathbb{R}^2$.

Theorem 7.4. Let $f : [a, b] \to \mathbb{R}$ be a continuous function and let $f(a) < t < f(b)$. Then $\exists$ at least one element $x \in [a, b]$ s.t. $f(x) = t$.

Proof. Since $[a, b]$ is connected, $f([a, b]) = A$ is connected by Theorem 7.3. Assume to the contrary that $\nexists$ any such $x \in [a, b]$.
    Let
$$\begin{aligned} U &= \{y \in \mathbb{R} | y < t\}, \\ V &= \{y \in \mathbb{R} | y > t\}. \end{aligned} \quad (7.9)$$

Then U and V are both non-empty open subsets of A and form a separation of A. This contradiction establishes the existence of the point $x \in [a, b]$. □

7.3 Connected Subspaces

A subset A of a topological space X is said to be *connected* if A, equipped with the relative topology, is connected. In other words, the subspace A is connected.

Example 1. Given any topological space X, the singleton $\{x\}$ is always a connected subspace of X.

Example 2. Consider $(\mathbb{R}, \tau_u)$ and the set $\mathbb{Q}$ of rationals. Then $\mathbb{Q}$ is a disconnected subspace of $\mathbb{R}$. To see this, let r be any irrational number. Then (U, V), where
$$\begin{aligned} U &= \{x \in \mathbb{Q} | x < r\}, \\ V &= \{x \in \mathbb{Q} | x > r\}, \end{aligned}$$

forms a separation of $\mathbb{Q}$ and hence $\mathbb{Q}$ is disconnected. Similarly, the set of integers, $\mathbb{Z}$, is disconnected because with the relative topology inherited from τ_u on $\mathbb{R}$, it is a discrete space and any discrete space (with more than one point) is always disconnected.

Example 3. In $(\mathbb{R}, \tau_c)$ both $\mathbb{Q}$ and $\mathbb{Z}$ are connected subspaces of $\mathbb{R}$ (verify!). Similarly, any infinite subset A of $\mathbb{R}$ is connected. However, if A is any finite subset of $\mathbb{R}$, then A is a disconnected subspace (verify!).

Example 4. Let $\mathbb{R}$ be endowed with the p–inclusion topology and A be any (finite or infinite) subset of $\mathbb{R}$. We first see what kind of relative or induced topology on A we have. There are only two possibilities: (i) either $p \in A$ or; (ii) $p \notin A$. (i) If $p \in A$, then the relative topology on A is precisely the p–inclusion topology, and as seen before (Section 7.1, Example 8), A is a connected subspace of $\mathbb{R}$. (ii) If $p \notin A$ then p cannot belong to any subset of A. Therefore, in this case, the relative topology on A is discrete, and hence A is disconnected.

Example 5. Let $X = \mathbb{R}^2$. Then any straight line in $\mathbb{R}^2$ is a connected subspace of $\mathbb{R}^2$, as seen before.

Example 6. $(a,b), [a,b], (a,\infty)$ and $(-\infty, a)$ are connected subspaces of $\mathbb{R}$ in $(\mathbb{R}, \tau_u)$. Similarly, the open square $(a,b) \times (a,b)$ and the closed square $[a,b] \times [a,b]$ are connected subspaces of $\mathbb{R}^2$.

Although quite reasonable from a visual standpoint, the criterion that a space is connected if it consists of solely "one piece" is often difficult to apply in practice. For a subspace A of a topological space X, there is another useful way of formulating the definition of connectedness.

Theorem 7.5. A subspace A of a toplogtical space X is connected iff for each pair of open (respectively closed) subsets U and V of X whose union contains A s.t. $A \cap U \neq \phi$ and $A \cap V \neq \phi$, then $A \cap U \cap V \neq \phi$.

Proof. We first prove sufficiency. Let A be a connected subspace of X and suppose that $A \cap U \cap V = \phi$. Then by definition of the relative topology of A, we easily get a contradiction to the fact that A is connected. In fact, since $A \cap U = A_1$, and $A \cap V = A_2$ are non-empty open subsets of A and $A_1 \cap A_2 = A \cap U \cap V = \phi$ by assumption, therefore A_1 and A_2 is a separation of A, i.e., A is disconnected contrary to our assumption. Thus $A \cap U \cap V \neq \phi$.

For necessity, assume that the subspace A is disconnected, then $\exists$ a pair of non-empty open subsets (A_1, A_2) of A s.t. $A = A_1 \cup A_2$ and $A_1 \cap A_2 = \phi$. Again by definition of the relative topology, $\exists$ a pair of non-empty open subsets (U, V) of X s.t. $A \cap U = A_1$ and $A \cap V = A_2$. Hence $A \cap U \cap V = \phi$. Therefore, if A is disconnected then $\exists$ a pair of open (respectively closed) subsets U, V of X s.t. $A \cap U \cap V = \phi$. This completes the proof. □

Application 7.3. In $(\mathbb{R}, \tau_u)$, $\mathbb{Z}$ is disconnected because if we take $U = \left(-\infty, \frac{1}{2}\right)$ and $V = \left(\frac{1}{2}, \infty\right)$, then U, V are open subsets of $\mathbb{R}$ s.t. $\mathbb{Z} \subset U \cup V$. Moreover, $\mathbb{Z} \cap U \neq \phi, \mathbb{Z} \cap V \neq \phi$ but $\mathbb{Z} \cap U \cap V = \phi$.

Application 7.4. Again in $(\mathbb{R}, \tau_u)$, $\mathbb{Q}$ is disconnected. For, if r is any irrational number then $U = (-\infty, r)$ and $V = (r, \infty)$ are both non-empty open subsets of $\mathbb{R}$ s.t. $\mathbb{Q} \subset U \cup V$ and $\mathbb{Q} \cap U \neq \phi, \mathbb{Q} \cap V \neq \phi$, however $\mathbb{Q} \cap U \cap V = \phi$.

Similarly the set $\mathbb{Q}'$ of irrational number is disconnected, because if x is any rational number then $A = (-\infty, x)$ and $B = (x, \infty)$ are both non-empty open subsets of $\mathbb{R}$ s.t. $\mathbb{Q}' \subset A \cup B$ and $\mathbb{Q}' \cap A \neq \phi, \mathbb{Q}' \cap B \neq \phi$, but $\mathbb{Q}' \cap A \cap B = \phi$.

Theorem 7.6. If the sets U and V form a separation of a topological space X, and if A is a connected subspace of X, then either A is fully contained in U or in V, i.e., $A \subset U$ or $A \subset V$.

Proof. Since U and V are both open in X, the sets $A \cap U = A_1$ and $A \cap V = A_2$ are open in A. Moreover, $A = A_1 \cup A_2$ and $A_1 \cap A_2 = \phi$. Now, if A_1 and A_2 were both non-empty, then they would constitute a separation of A. Therefore, by the connectedness of A we must have either $A_1 = \phi$ or $A_2 = \phi$. Hence A must lie entirely in U or in V. □

Theorem 7.7. A topological space X is connected iff for each pair of points $x, y \in X$ there exists a connected subspace A containing them.

Proof. Necessity is obvious since if X is connected, then for each pair of points of X, the set X itself can serve the purpose.

Conversely, assume that the condition of the theorem holds and X is disconnected. Then $\exists$ a pair (U, V) of non-empty disjoint open subsets of X s.t. $X = U \cup V$. Let x, y be any pair of points of X, then by assumption there exists a connected subspace A of X containing both x and y. Assume that $x \in U$ and $y \in V$. We can do so because $U \cap V = \phi$. By virtue of Theorem 7.6, either A lies entirely within U or V. This is an obvious contradiction of the fact that $A \cap U \neq \phi$ and $A \cap V \neq \phi$. Hence our assumption is wrong, and thus X is connected. This completes the proof. □

7.4 Other Characterizations of Connected Subspaces

We have other useful characterizations of the connectedness of subspaces.

Theorem 7.8. Let A be a dense subset of a topological space X. If A is connected, then so is X.

Proof. Let A be a connected subspace of X s.t. $\overline{A} = X$. Assume, to the contrary, that X is disconnected. Then $\exists$ non-empty closed subsets F_1 and F_2 of X s.t. $F_1 \cup F_2 = X$ and $F_1 \cap F_2 = \phi$. Since A is connected, by virtue of Theorem 7.6, A lies entirely either in F_1 or in F_2. Let $A \subset F_1$ (say), then $\overline{A} \subset \overline{F_1} = F_1$ because F_1 is closed in X. Hence, it follows that $\overline{A} = X \subset F_1$, which contradicts the fact that $F_2 \neq \phi$. Thus X is connected. □

Application 7.5. For $(\mathbb{R}, \tau_c)$ $\mathbb{N}$ is dense in $\mathbb{R}$. Moreover, $\mathbb{N}$ is connected with respect to the relative topology. (Note that this is precisely the co-finite topology on $\mathbb{N}$). Hence $\mathbb{R}$ is also connected.

Application 7.6. Let $X = [0, 1]$ (or any $[a, b]$) and $A = (0, 1)$. Then $\overline{A} = X$. Since $(0, 1)$ is connected, so is $[0, 1]$.

Note that there may exist dense subsets of a connected space which are not connected. For instance, in $(\mathbb{R}, \tau_u)$ $\overline{\mathbb{Q}} = \mathbb{R}$ and $\mathbb{R}$ is connected but $\mathbb{Q}$ is not connected, as we have seen before.

Theorem 7.9. If A is a dense and connected subspace of a topological space X, then every set B s.t. $A \subseteq B \subseteq \overline{A}$ is connected. In particular, $\overline{A}$ is also connected.

Proof. Let A be connected and let $A \subseteq B \subseteq \overline{A}$. Suppose that B is disconnected, then $\exists$ non-empty closed subsets F_1 and F_2 of the subspace B s.t. $B = F_1 \cup F_2$ and $F_1 \cap F_2 = \phi$. Note that A is a connected subspace of the subspace B. By Theorem 7.3, A must lie entirely in F_1 or in F_2. Let $A \subseteq F_1$ (say), then $\overline{A} \subseteq \overline{F_1} = F_1$ because F_1 is closed. Since F_1 and F_2 are disjoint and A is a dense subset of the subspace B, hence $B \subseteq F_1$. This contradicts the fact that F_2 is a non-empty subset of B. Hence B is connected. □

It is an important property of the set of real numbers that it is a connected space relative to the usual topology. This fact follows from the next theorem.

Theorem 7.10. Every interval of $\mathbb{R}$ is connected.

Proof. Let $X = (a, b)$. Assume to the contrary that X is disconnected and let $X = A \cup B$, where A, B are non-empty closed disjoint subsets of X. Choose $x \in A$ and $y \in B$ s.t. $x \neq y$, and suppose (without loss of generality) that $x < y$. Since X is an interval, $[x, y] \subseteq X$ and each point in $[x, y]$ is either in A or in B, but not in both. Now define z by, $z = \sup([x, y] \cap A)$. Then $x \leq z \leq y$ and hence $z \in X$. Since A is closed in X, we must have $z \in A$ by the definition of z, and thus $z < y$. Again by the definition of the supremum of z, $z + \epsilon$ is in B for every $\epsilon > 0$ s.t. $z + \epsilon \leq y$. Since B is closed in X, it follows that $z \in B$. Thus, $z \in A \cap B$ which is an obvious contradiction to our assumption that these sets are disjoint. Hence X is connected. □

See Example 6 in Section 7.3 for an application of above result.

Corollary 7.10.1. $\mathbb{R}$ is connected with respect to the usual topology.

Proof. Since (a, b) is homeomorphic to $\mathbb{R}$, the result follows from the corollary to Theorem 7.3 together with Theorem 7.10. □

Corollary 7.10.2. The closed interval $[a, b]$ is connected.

Proof. The result follows from Theorem 7.9. □

The subsets A, B of a topological space X are called *separated* if

$$\overline{A} \cap B = \phi = A \cap \overline{B}. \tag{7.10}$$

Note that two disjoint subsets are separated if and only if neither of them contains an accumulation point of the other. In particular, any two closed and disjoint subsets A, B of X are separated. Similarly, any two open and disjoint subsets of X are separated. For instance, if $X = \mathbb{R}$ equipped with the usual topology, and if $A = (0, 1)$ and $B = (1, 2)$ then

$$\overline{A} \cap B = [0, 1] \cap (1, 2) = \phi = (0, 1) \cap [1, 2] = A \cap \overline{B}. \tag{7.11}$$

Also note that if the subsets A and B of X are separated, and if $A_1 \subseteq A, B_1 \subseteq B$, then A_1 and B_1 are also separated.

Example 1. In $(\mathbb{R}, \tau_u)$, $\mathbb{Q}$ and $\mathbb{Q}'$ are not separated because $\overline{\mathbb{Q}} \cap \mathbb{Q}' = \mathbb{Q}'$ and $\mathbb{Q} \cap \overline{\mathbb{Q}'} = \mathbb{Q}$.

Example 2. Any two infinite disjoint subsets of $(\mathbb{R}, \tau_c)$ are not separated. In fact, if A and B are any two infinite disjoint subsets of $\mathbb{R}$, then

$$\overline{A} \cap B = \mathbb{R} \cap B = B, \text{ and } A \cap \overline{B} = A \cap \mathbb{R} = A. \tag{7.12}$$

We present now a final useful characterization of connected subsets of X.

Theorem 7.11. A subspace of a topological space X is connected iff for each pair of separated sets U and V with $A = U \cup V$, we have either $U = \phi$ or $V = \phi$.

Proof. The proof of this theorem is left as a simple exercise. □

7.5 Connectedness of Families of Sets

The intersection of two connected sets may fail to be connected. A simple example, shown in Figure 7.2, is provided by a circle with two points removed from it. Consider the circle, S, and the two subsets $U_1 = \mathbb{S}^1 \smallsetminus \{(1,0)\}$ and $U_2 = \mathbb{S}^1 \smallsetminus \{(-1,0)\}$. Clearly U_1 and U_2 are connected subsets of $V = U_1 \cap U_2 = \mathbb{S}^1 \smallsetminus \{(1,0),(-1,0)\} = A \cup B$ and $A \cap B = \phi$, where A and B are the sets indicated in the figure. Thus V is a disconnected space. The union of two connected sets may also fail to be connected. For instance A and B in the above example. Or for instance, in a co-finite space the union of two distinct singletons is disconnected. Similarly, in $(\mathbb{R}, \tau_u)$ the sets $(-\infty, 0)$ and $(0, \infty)$ are connected but their union $(-\infty, 0) \cup (0, \infty)$ is disconnected. However, under certain conditions, arbitrary unions of connected sets are connected.

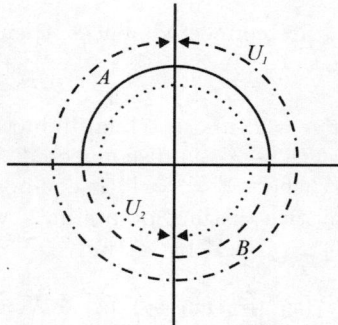

Figure 7.2: The intersection of two connected sets need not be connected. For example take $U_1 = \mathbb{S}^1 \smallsetminus \{(1,0)\}$ (dot-dashed) and $U_2 = \mathbb{S}^1 \smallsetminus \{(-1,0)\}$ (dotted). Then $U_1 \cap U_2 = A \cup B$ where A is the upper open semi-circle (solid) and B the lower open semi-circle (dashed).

Theorem 7.12. *The union of a family $\{A_\alpha\}_{\alpha \in I}$, of connected sets whose intersection is non-empty is connected.*

Proof. Let $\{A_\alpha\}_{\alpha \in I}$ be a family of connected subsets of a topological space X s.t. $\bigcap_{\alpha \in I} A_\alpha \neq \phi$. Assume that $x \in \bigcap_{\alpha \in I} A_\alpha$. We have to prove that $A = \bigcup_{\alpha \in I} A_\alpha$ is connected. Assume to the contrary that A is disconnected. Then, by virtue of Theorem 7.5, $\exists$ two open subsets U and V of X such that $A \cap U \neq \phi, A \cap V \neq \phi$ and $A \subset U \cup V$ and $A \cap U \cap V = \phi$. On the one hand, the point x belongs to one of the sets U, V, say $x \in U$. On the other hand, one of the sets, say A_α, meets V. Therefore we have

$$A_\alpha \subset U \cup V, \; A_\alpha \cap U \cap V = \phi, \tag{7.13}$$

and

$$A_\alpha \cap U \neq \phi, \; A_\alpha \cap V \neq \phi. \tag{7.14}$$

Hence A_α is disconnected, by Theorem 7.5, which is an obvious contradiction to our hypothesis. Thus A is connected. □

Application 7.7. In $(\mathbb{R}, \tau_u)$, the set $(0, 1) \cup (1, 2)$ is disconnected because $(0, 1) \cap (1, 2) = \phi$ although both $(0, 1)$ and $(1, 2)$ are connected.

Application 7.8. Consider the family of sets $A_n = (-n, n)$, where $n \in \mathbb{N}$. Then $\bigcap_{n \in \mathbb{N}} A_n \neq \phi$. Hence $\bigcup_{n \in \mathbb{N}} (-n, n) = \mathbb{R}$ is connected.

Theorem 7.13. Let $\{A_n\}_{n \geq 0}$ be an infinite sequence of connected subsets of a topological space X s.t. $A_{n+1} \cap A_n \neq \phi, \forall n \geq 0$. Then the union $\bigcup_{n=0}^{\infty} A_n$ is connected.

Proof. By induction on n and the use of Theorem 7.5, we see immediately that the set $B_n = \bigcup_{i=0}^{n} A_i$ is connected $\forall n$. Moreover, $\bigcap_{n=0}^{\infty} B_n \neq \phi$. Hence, by Theorem 7.12, $\bigcup_{n=0}^{\infty} B_n = \bigcup_{n=0}^{\infty} A_n$ is connected. □

Now we prove that the product of finitely many connected spaces is connected. This is of crucial importance in constructing product spaces.

Theorem 7.14. If $X_1, X_2, \cdots, X_n$ are connected spaces, then the product space $X = \prod_{i=1}^{n} X_i$ is also connected.

Proof. We prove the theorem for $n = 2$, and a trivial inductive argument will yield the more general result. Assume that $X = X_1 \times X_2$ is disconnected, then by Theorem 7.2, there is a continuous function from X onto Y, where $Y = \{0, 1\}$ is equipped with the discrete topology. Let $a = (a_1, a_2)$ and $b = (b_1, b_2)$ be any two distinct points in X with $f(a) = 0$ and $f(b) = 1$. Now define function $f_1 : X_1 \to Y$ and $f_2 : X_2 \to Y$ by setting

$$f_1(x_1) = f(x_1, b_2) \; \forall x_1 \in X_1, \tag{7.15}$$

$$f_2(x_2) = f(a_1, x_2) \; \forall x_2 \in X_2. \tag{7.16}$$

On the one hand, it is easily seen that f_1 and f_2 are continuous, and since X_1 and X_2 are connected, we conclude that f_1 and f_2 must be constant functions by virtue of Corollary 7.2.1. Since $f_1(b_1) = f(b_1, b_2) = 1$, it follows that f_1 must be identically unity. Since $f_2(a_2) = f(a_1, a_2) = 0$, it follows that f_2 is identically zero. On the other hand, we have that

$$f_1(a_1) = f_2(b_2) = f(a_1, b_2), \tag{7.17}$$

an obvious contradiction. Thus we conclude that X is connected. □

See Example 6 in Section 7.3 for an application of this result. We can now take a product of three spaces X_1, X_2, X_3 as a product of $(X_1 \times X_2)$ with X_3 and use the above result twice. This procedure can be carried out for any n.

Corollary 7.14.1.

(i) The space $\mathbb{R}$ is connected;

(ii) $\prod_{i=1}^{n} (a_i, b_i)$ is connected.

In particular, the open square $(a, b) \times (a, b)$ and the closed square $[a, b] \times [a, b]$ are connected. Similarly, the torus $\mathbb{S}^1 \times \mathbb{S}^1$ is connected.

Exercise 7.5. Show that if the product $\prod_{i=1}^{n} X_i$ is non-empty and connected, then each X_i is connected. [Hint: Consider the case when $n = 2$, and use the projection mapping $p_i : X_1 \times X_2 \to X_i \; (i = 1, 2)$.]

7.6 Path Connectedness

In this section, we introduce a somewhat different concept of connectedness, known as path connectedness, which is frequently useful in Homotopy theory and in Geometry. We also study some of its basic properties.

Let x and y be any two points of a topological space X. A path from x to y in X is a continuous function $f : [0, 1] \to X$ s.t. $f(0) = x$ and $f(1) = y$. The points x and y are called the *initial point* and *end point* or *terminal point* of f respectively, see Figure 7.3. If X is a geometrical space, f is called a *curve* in X.

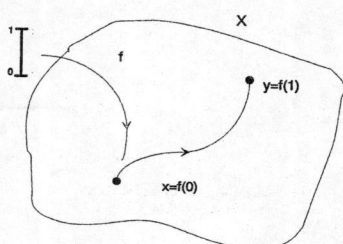

Figure 7.3: A path from x to y is a mapping from the closed unit interval I into X which is continuous and $f(0) = x, f(1) = y$.

A topological space X is said to be *path connected* if every pair of distinct points of X can be joined by a path in X, i.e., from every $x, y \in X, \exists$ a continuous function $f : [0, 1] \to X$ s.t. $f(0) = x$ and $f(1) = y$.

Example 1. $(\mathbb{R}, \tau_u)$ is path connected, for if $\mathbf{x}$ and $\mathbf{y}$ are in $\mathbb{R}$, the straight line path $f : [0, 1] \to \mathbb{R}$ defined by $f(t) = (1 - t)x + ty$, $t \in [0, 1]$ lies in $\mathbb{R}$.

Example 2. $\mathbb{R}^2$ with respect to the Euclidean topology is path connected, with the same argument as given above, replacing x and y by the vectors $\mathbf{x}$ and $\mathbf{y}$.

Example 3. The closed unit sphere in $\mathbb{R}^n$,

$$\overline{S_1(0)} = \left\{ \mathbf{x} \in \mathbb{R}^n \,|\, x_1^2 + x_2^2 + \cdots + x_n^2 \leq 1 \right\}, \tag{7.18}$$

is path connected. In fact, given any $\mathbf{x}, \mathbf{y} \in \overline{S_1(0)}$ the straight-line path $f : [0, 1] \to \mathbb{R}^n$ defined by

$$f(t) = (1 - t)\mathbf{x} + t\mathbf{y}, \tag{7.19}$$

lies in $\overline{S_1(0)}$ because

$$\|f(t)\| \leq (1 - t)\|\mathbf{x}\| + t\|\mathbf{y}\| \leq 1, \tag{7.20}$$

where

$$\|\mathbf{x}\| = \left(x_1^2 + x_2^2 + \cdots + x_n^2 \right)^{\frac{1}{2}} \leq 1, \quad \forall\, \mathbf{x} \in \overline{S_1(0)}. \tag{7.21}$$

A similar argument shows that each open unit sphere in $\mathbb{R}^n$ is path connected.

Example 4. Consider $\mathbb{R}^n \smallsetminus \{O\}$, where O is the origin in $\mathbb{R}^n$ (for $n \geq 2$). Then $\mathbb{R}^n \smallsetminus \{O\}$ is path connected. For, if $P, Q \in \mathbb{R}^n \smallsetminus \{O\}$ then we can join $\mathbf{x}$ and $\mathbf{y}$ by the straight-line path between them such that the path does not go through the origin. If it does, we can choose a point R not on the line joining P and Q and take the broken-line path from P to R, and then from R to Q, see Figure 7.4. Clearly $\mathbb{R}^n \smallsetminus \{O\}$ is also connected.

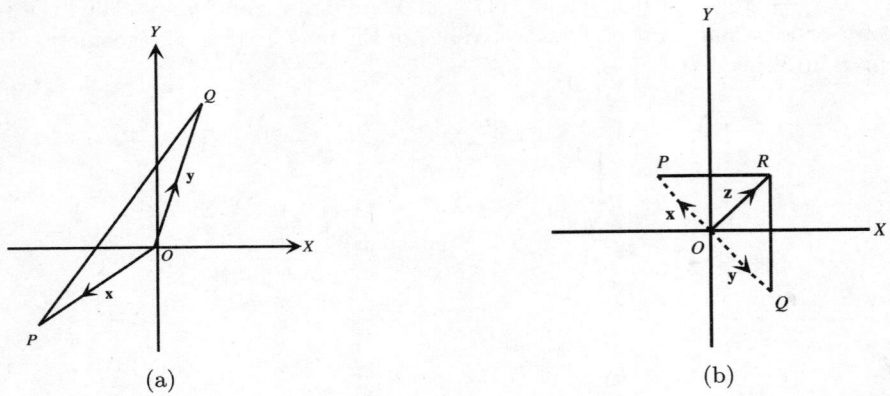

(a)   (b)

Figure 7.4: Two points in $\mathbb{R}^n \smallsetminus \{O\}$ are always path connected. Denote the position vectors of P and Q by $\mathbf{x}$ and $\mathbf{y}$. If (a) $\mathbf{x} \not\parallel \mathbf{y}$ then we can take the path $\overrightarrow{PQ} = \mathbf{y} - \mathbf{x}$ joining them. If (b) $\mathbf{x} \parallel \mathbf{y}$, choose a point R with position vector $\mathbf{z} \perp \mathbf{x}$ or $\mathbf{y}$ and take the path as $\overrightarrow{PR} = \mathbf{z} - \mathbf{x}$ and then $\overrightarrow{RQ} = \mathbf{y} - \mathbf{z}$ to connect P to Q via R.

Example 5. Consider the surface of the open unit sphere, denoted by $\mathbb{S}^{n-1}$, defined as

$$\mathbb{S}^{n-1} = \left\{ \mathbf{x} \in \mathbb{R}^n \mid \|\mathbf{x}\| = \left(x_1^2 + x_2^2 + \cdots + x_n^2\right)^{\frac{1}{2}} = 1 \right\}, \ n \geq 2. \tag{7.22}$$

Then $\mathbb{S}^{n-1}$ is path connected, since the map $f : \mathbb{R}^n \smallsetminus \{0\} \to \mathbb{S}^{n-1}$ defined by $f(x) = \frac{x}{\|x\|}$ is continuous and surjective.

In particular the circle,

$$\mathbb{S}^1 = \left\{ (x_1, x_2) \in \mathbb{R}^2 \mid x_1^2 + x_2^2 = 1 \right\}, \tag{7.23}$$

is path connected.

Example 6. The space $X = Y \cup Z$ with the relative topology inherited from $\mathbb{R}^2$, where

$$Y = \left\{ \left(x, \sin \frac{1}{x}\right) \in \mathbb{R}^2 \mid 0 < x < \frac{1}{\pi} \right\}, \tag{7.24}$$

$$Z = \left\{ (x, y) \in \mathbb{R}^2 \mid x = 0 \text{ and } -1 \leq y \leq 1 \right\}. \tag{7.25}$$

is frequently referred to as the topologist's sine curve, see Figure 7.5.

It follows immediately from Theorem 7.12 that X is connected. However, X is not path connected as there is no path connecting a point with positive x to one with negative x. To see

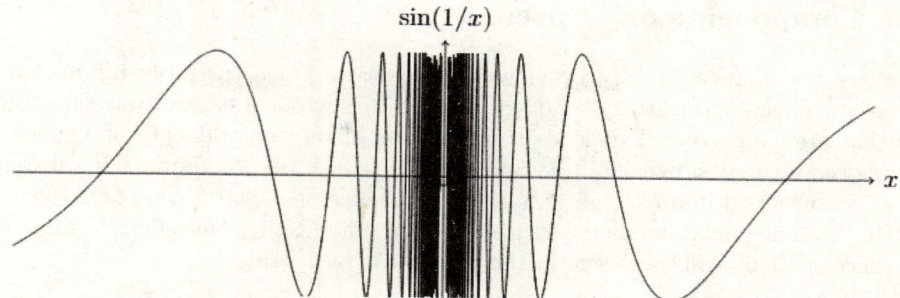

Figure 7.5: The topologist's sine curve, is clearly connected, however it is not path connected.

this, consider the points $\left(\frac{1}{\pi}, 0\right)$ and $(0,0)$. These points cannot be joined by a path in X. Assume to the contrary that there *is* a path $f : [0,1] \to X$ with $f(0) = \left(\frac{1}{\pi}, 0\right)$ and $f(1) = (0,0)$. Since $f([0,1])$ is connected, every point on the sine curve (for $0 < x < \frac{1}{\pi}$) must be included in the range of f. Thus we may select a sequence of points in $[0,1]$, $x_1 < x_2 < \cdots < x_n < \cdots$ s.t. the sequence $\{x_n\}$ converges to 1 and the second coordinate of $f(x_n)$ is 1 if n is odd and is -1 if n is even. This, however, is absurd, since as $\{x_n\}$ converges to 1, the sequence $\{f(x_n)\}$ attempts to simultaneously approach both $(0,-1)$ and $(0,1)$ which is impossible. Thus we conclude that X is not path connected. This demonstrates the difference between connectedness and path connectedness.

We now show that if the space is path connected, then it is necessarily connected.

Theorem 7.15. *Every path connected space is connected.*

Proof. Let X be a path connected space and let x, y be any pair of distinct points of X. Since X is path connected, $\exists\, f : [0,1] \to X$ s.t. $f(0) = x, f(1) = y$. Further, since $[0,1]$ is connected, the trajectory of the path $f([0,1])$ is also a connected subset of X (by Theorem 7.3), containing the points x and y. Hence X is connected by virtue of Theorem 7.7. $\square$

The converse of this statement is not true as shown above in Example 6. Like connectedness, path connectedness is also a topological property, as we see in the following theorem.

Theorem 7.16. *Path connectedness is a topological property.*

Proof. Let X be path connected and $g : X \to Y$ a homeomorphism of X onto another topological space Y. We have to show that Y is also path connected. To see this, let y_1 and y_2 be any pair of distinct points of Y. Then $\exists\, x_1, x_2 \in X$ s.t. $g(x_1) = y_1, g(x_2) = y_2$. Since X is path connected, $\exists\, f : [0,1] \to X$ s.t. $f(0) = x_1, f(1) = x_2$. Then $g \circ f : [0,1] \to Y$ is the desired path joining y_1 and y_2. Hence Y is also path connected. $\square$

Theorem 7.17. *Let $X_1, X_2, \cdots, X_n$ be path connected spaces. Then $X = \prod_{i=1}^{n} X_i$ is also connected.*

Proof. The proof is left as a simple exercise to the reader. $\square$

7.7 The Components of a Space

Spaces that are not connected may be viewed as consisting of a (possibly infinite) number of connected pieces. Given an arbitrary topological space X, there is a natural way of breaking it up into pieces that are connected. This leads to the notion of the "component" of a space.

A maximal connected subspace of a topological space X, i.e., a connected subspace which is not properly contained in any larger connected subspace, is called a *component*, or *connected component*, of X. Since maximal elements need not be unique, there may be several components in a given space X. This will be shown in the examples given below.

Example 1. Let X be any connected space. For definiteness think of X as $(\mathbb{R}, \tau_u)$. Then, obviously, the only component of X is X itself. As the other extreme, if X is any discrete space, then each singleton $\{x\}$ is a component of X.

Example 2. $X = [-2, 0) \cup (0, 2]$, equipped with the relative topology inherited from the usual topology of $\mathbb{R}$, has components [-2,0) and (0,2].

Example 3. Since the circle $\mathbb{S}^1$ is path connected, and hence connected, there is only one component of $\mathbb{S}^1$, namely $\mathbb{S}^1$ itself.

Example 4. $\mathbb{Q}$ is not a connected subset of $\mathbb{R}$ (with respect to the usual topology). The components of $\mathbb{Q}$ are the singletons $\{x\}, x \in \mathbb{Q}$. (Verify!)

Example 5. $\mathbb{R} \smallsetminus \mathbb{Q}$ has countably infinitely many components $\mathbb{R}$ for each element x of $\mathbb{Q}$ and is a disconnected space.

Example 6. Let X be the subspace of $\mathbb{R}^2$ consisting of equally spaced horizontal line segments or vertical line segments, as shown in Figure 7.6.

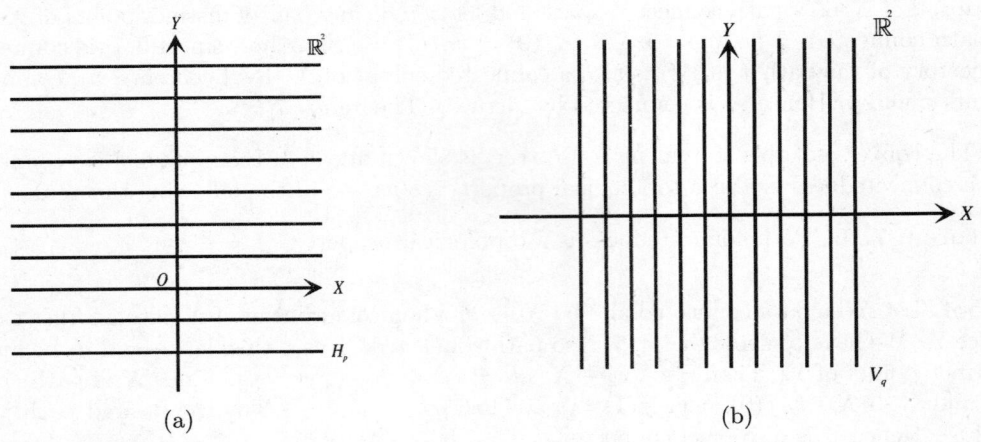

Figure 7.6: (a) The horizontal line segments H_p ($p = -n, \cdots, n$) are not connected and so are components of X. (b) Similarly, the vertical line segments V_q ($q = -m, \cdots, m$) are components of X.

Then each vertical or horizontal line segment is a component of X. (Verify!)

Theorem 7.18. For a topological space X:

(i) each $x \in X$ is contained in exactly one component;

(ii) distinct components of X are disjoint;

(iii) each component of X is closed.

Proof.

(i) Let x be any point of X. Consider the collection $\{C_\alpha\}_{\alpha \in I}$ of all connected subspaces of X which contain x. Since $\{x\}$ is one such set, the collection is non-empty. Further, since $\bigcup_{\alpha \in I} C_\alpha = C_x$ is a connected subspace of X by Theorem 7.12. Because of its construction, C_x is the maximal connected subspace of X containing x. Therefore, C_x is a component of X containing the point x.

(ii) Let C_{x_1} and C_{x_2} be any two distinct components of the points x_1 and x_2. We claim that $C_{x_1} \cap C_{x_2} = \phi$. For, if $x_0 \in C_{x_1} \cap C_{x_2}$, then by virtue of Theorem 7.12, their union $C_{x_1} \cup C_{x_2} = C$ is also connected. Therefore, $C \subseteq C_{x_i} (i = 1, 2)$ by the maximality of C_{x_1} and C_{x_2}, and thus we conclude that $C = C_{x_1} = C_{x_2}$. Hence the components are either disjoint or they coincide.

(iii) Let C_x be any component of X containing the point x. Then, by virtue of Theorem 7.9, $\overline{C}_x$ is also connected. Because of the maximality of C_x, we have $\overline{C}_x \subseteq C_x$, i.e., $\overline{C}_x = C_x$. Hence every component, C_x is closed.

□

Corollary 7.18.1. Every connected subspace of a topological space X is contained in a component of X which contains all of its points. Further, $\bigcup_{x \in X} C_x = X$.

Proof. See the construction in Theorem 7.18(i). □

Here is another simple characterization of connected spaces.

Theorem 7.19. A topological space X is connected iff there is exactly one component, namely X itself.

Proof. Assume that X is connected, then X is the only component of X because it is the maximal connected subspace of X. Conversely, if X has exactly one component, namely X itself, then X is obviously connected. □

We just proved that every component is a closed set. It is reasonable to ask whether it is open. The answer is 'no', as seen from Example 4 above. However, if a connected subset of X is both open and closed, then it is necessarily a component of X as shown below. Here is another example demonstrating that components need not be open.

Example 7. In $\mathbb{R}^2$, for each $n \in \mathbb{Z}^+$, let

$$H_n = \left\{\left(x, \frac{1}{n}\right) | 0 \le x \le 1\right\},$$
$$H_0 = \{(x, 0) | 0 \le x \le 1\}. \tag{7.26}$$

Then the components of $X = H_0 \cup (\bigcup_{n=1}^{\infty} H_n)$ are precisely the sets $H_0, H_1, \cdots,$ $H_n, \cdots$ as is clear from Figure 7.6 and the example associated with it. Although, by Theorem 7.18(iii), H_0 must be closed, H_0 is clearly not open in X, and thus we see that components need not be open.

Here is another necessary condition under which a connected subset is a component.

Theorem 7.20. If A is a connected subspace of a topological space X, which is both open and closed, then A is a component of X.

Proof. First note, by virtue of Corollary 7.18.1, every connected subset is contained in some component of X. Let A be a connected subspace of X which is both open and closed. Then it is contained in some component C of X. We show that $A = C$. Assume that A is properly contained in C, i.e., $A \subset C$, then

$$(C \cap A) \cup (C \cap A') = C. \tag{7.27}$$

That is, $(C \cap A, C \cap A')$ forms a separation of C which contradicts the fact that C, being a component of X, is connected. Hence $A = C$. This completes the proof. $\square$

If we drop the condition of connectedness in the preceding theorem, we get the following result:

Theorem 7.21. Let A be any non-empty open and closed subset of a space X. Then it is a union of components of X.

Proof. We first show that for every $a \in A$, the component $C_a \subseteq A$. Since $C_a \cap A$ is both open and closed in the subspace C_a, and C_a is connected, it must follow from

$$(C_a \cap A) \cup (C_a \cap A') = C_a, \tag{7.28}$$

that either $C_a \cap A = \phi$ or $C_a \cap A' = \phi$ by virtue of Theorem 7.1. Since A is non-empty, $C_a \cap A \ne \phi$ and hence $C_a \cap A = C_a$, i.e., $C_a \subseteq A$. Therefore, $\bigcup_{a \in A} C_a \subseteq A$. The reverse inequality being trivial, we conclude that $A = \bigcup_{a \in A} C_a$. This completes the proof. $\square$

Theorem 7.22. Let $f : X \to Y$ be a continuous function of a topological space X into Y. If B is any component of Y, then $f^{-1}(B)$ is a union of components of X.

Proof. Let $f^{-1}(B) = A, a \in A$, and let C_a be a component of X containing the point a. Then $f(C_a)$ is connected and also $f(C_a) \cap B \ne \phi$. Hence, by Corollary 7.18.1, $f(C_a) \subseteq B$. This implies that

$$C_a \subseteq f^{-1}(B) = A. \tag{7.29}$$

Thus, $A = \bigcup_{a \in A} C_a$. This completes the proof. $\square$

There is another way of defining the component. Define an equivalence relation on the given space X by setting $x \sim y$ if there is a connected subset of X containing both x and y. Symmetry and reflexivity of the relation are obvious. Now, we consider transitivity. If $x \sim y$ and $y \sim z$ then $\exists$ connected $A, B \subseteq X$ s.t. $x, y \in A$ and $y, z \in B$. Since $y \in A \cap B$, $A \cup B$ is connected and contains x and z i.e., $x \sim z$. Hence transitivity of the relation also holds.

The equivalence class
$$C_x = \{y \in X | x \sim y\}, \tag{7.30}$$
of the point x is precisely the component of X containing x. It is obvious that $\bigcup_{x \in X} = X$, and any two equivalence classes are either disjoint or the same. Let $y \in C_x$, then $\exists$ a connected subset A_y s.t. $x, y \in A_y$. Note that $A_y \subseteq C_x$. Hence $\bigcup_{y \in C_x} A_y = C_x$ is connected because $x \in \bigcap_{y \in C_x} A_y$. Thus C_x is a maximal connected subset of X and hence it is a component of X.

7.8 Total Disconnectedness

We now define another notion of a "totally disconnected space". A topological space X is said to be *totally disconnected* if its only connected subspaces are singleton sets. In other words, X is totally disconnected if the component of every point, $x \in X$, is just the singleton, $\{x\}$. Totally disconnected spaces are of considerable significance in several parts of Topology, specifically in dimension theory and in the representation theory for Boolean algebras.

Example 1. $\mathbb{Q}$ is a totally disconnected subspace of $(\mathbb{R}, \tau_u)$. For, if A is any connected subset of $\mathbb{Q}$ consisting of at least two points x and y, take an irrational number, r, lying between x and y. Then $[r, \infty) \cap A$ is a non-empty open and closed set in the subspace A. Hence A is not connected, contradicting our assumption.

Example 2. It can similarly be shown that the set of irrational numbers is totally disconnected.

Example 3. The space $\mathbb{R} \setminus \{0\}$ is disconnected as seen before, however it is not totally disconnected.

Example 4. Every finite Hausdorff space is totally disconnected (Verify!).

Example 5. Every discrete space is totally disconnected. Note that a totally disconnected space need not be a discrete space, see Example 1 above.

Example 6. The Cantor set (obtained by repeatedly deleting the central one thirds of an initial line segment, see Figure 7.7) is totally disconnected. (Verify!)

Theorem 7.23. A topological space is totally disconnected if for every pair of distinct points x and y in X, $\exists$ a separation $X = A \cup B$ with $x \in A$ and $y \in B$.

Proof. The proof is left as a simple exercise. $\square$

Note that every totally disconnected space is Hausdorff.
Here is another characterization of totally disconnected spaces.

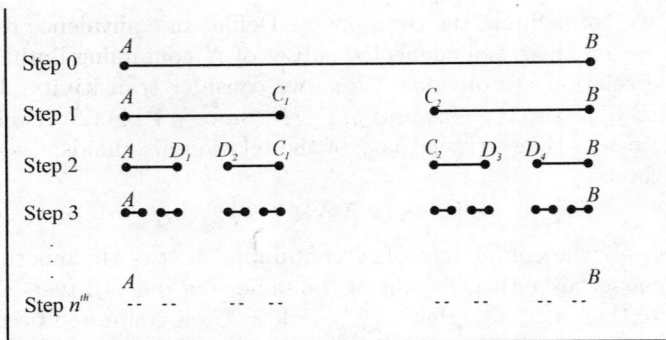

Figure 7.7: Construction of the Cantor set. The central third of AB is deleted to give $2 = 2^1$ equal segments. Then the central one third of each segment is deleted to give $4 = 2^2$ equal segments. Then the central one third of each segment is deleted to give $8 = 2^3$ segments. Thus at the nth step there are 2^n segments. This procedure is carried on infinitely.

Theorem 7.24. Let X be a Hausdorff space and let β be a bases of X whose sets are also closed in X, then X is totally disconnected.

Proof. We have only to show that the components of X are singletons. Let $x \in X$ and let C_x be the component of X containing x. Assume to the contrary that $\exists\, y \in C_x$ s.t. $x \neq y$. Since X is Hausdorff, x has a nbd U which does not contain y. By our assumption, $\exists$ a $basic$ open set (i.e., an open set that is also member of a base) G in β which is also closed such that $x \in G \subseteq (U)$. Hence

$$(G \cap C_x) \cup (G' \cap C_x) = C_x, \qquad (7.31)$$

i.e., C_x has a separation, and therefore it is not connected. This contradiction shows that our assumption is wrong, and thus $C_x = \{x\}$. Therefore X is totally disconnected. This completes the proof. $\square$

Under the additional assumption of compactness on X, the reverse implication also holds, as stated in the following theorem.

Theorem 7.25. Let X be a compact Hausdorff space. Then X is totally disconnected iff it has an open base β whose sets are also closed.

Proof. Necessity has already been proved in Theorem 7.24. For sufficiency, assume that X is totally disconnected. We need to prove that the class of all subsets of X which are both open and closed forms an open base for X. Let $x \in X$ and G be an open set $s.t.$ $x \in G$. We have only to find a set U which is both open and closed $s.t.$ $x \in U \subseteq G$. If $G = X$, we may take $U = X$. Therefore, assume that $G \neq X$. Since X is compact, G' is also compact, being the closed subspace of X. Since by hypothesis X is totally disconnected, for each point y in G' $\exists$ a set V_y which is both open and closed and contains y but not x. Therefore, $\{V_y | y \in G'\}$ forms an open covering of G'. By the compactness of G' $\exists$ a finite subcover of G', denoted by $\{V_1, V_2, \cdots, V_n\}$. Let $V = \bigcup_{i=1}^{n} V_i$. Then, obviously $G' \subseteq V$ and $x \notin V$. Since each V_i is both open and closed, V is also both open and closed. If we now take $U = V'$, then U satisfies our requirement, i.e., $x \in U \subseteq G$. This completes the proof. $\square$

7.9 Locally Connected Spaces

As we have seen before, connectivity is a very useful and important property which a given space may or may not possess. It may happen that the space satisfies this property locally i.e., each point has small neighborhoods that are connected. Such spaces are known as "locally connected spaces" whose formal definition is given as follows. A space X is called *locally connected* if $\forall x \in X$ and each open set $U \ni x, \exists$ a connected open set V s.t. $x \in V \subseteq U$. This definition is equivalent to saying that X is locally connected if there is a base β for X consisting of connected sets.

There is another way to define local connectedness which is generally used in the literature. A space X is locally connected if the components of every open subsets of X are open in X.

We shall establish that these notions are essentially equivalent. Local connectedness and connectedness of a space need not be related to one another. We shall show later that connectedness does not guarantee local connectedness and vice versa. In general, a space may possess one or both of these properties, or neither. Thus, we cannot regard local connectedness as a generalization or a special case of connectedness.

Example 1. Let X be any non-empty set with the indiscrete topology. Then X is locally connected. Note that it is also connected. If X is equipped with the discrete topology, then X is locally connected, because for every $x \in X$ and for every *nbd* U of x, there is a connected *nbd* V, namely $\{x\}$, of x contained in U. However, X is obviously not connected.

Example 2. In $(\mathbb{R}, \tau_u)$ $\forall x \in \mathbb{R}$ and for every *nbd* U of x $\exists$ a connected *nbd*, namely the open interval $V = (x - \epsilon, x + \epsilon), s.t.\ x \in V \subset U$.

Similarly, each interval is also locally connected. These spaces are also connected.

Example 3. Any non-empty infinite topological space (X, τ_c) is locally connected. In fact, $\forall x \in X$ and for every neighborhood V of x, namely the open set containing $x, s.t.\ x \in V \subseteq U$. Note that in the co-finite space, a subset V is connected iff it is either the singleton, $\{x\}$, or infinite.

Example 4. $X = [0,1) \cup (1,2]$ with the relative topology inherited from τ_u on $\mathbb{R}$ is not connected but it is locally connected.

Example 5. The subspace $(\mathbb{R}, \tau_U)$ of $(\mathbb{R}, \tau_U)$ is neither connected nor locally connected.

Now we give an example of a space which is connected but not locally connected.

Example 6. Let $A = \left\{\frac{1}{n} | n \in \mathbb{N}\right\}$ and define

$$X = ([0,1] \times \{0\}) \cup (A \times [0,1]) \cup (\{0\} \times [0,1]) \tag{7.32}$$

as shown in Figure 7.8(a). The space X together with the relative topology inherited from $\mathbb{R}^2$ is called the "Comb Space". It can be shown easily that X is connected. We claim that X is not locally connected. Assume that X is locally connected. As we shall see later, every open subspace of a locally connected space is itself locally connected, therefore the open subspace

$$U = \left\{(x,y) \in X | y < \frac{1}{2}\right\}, \tag{7.33}$$

is locally connected (see Figure 7.8(b)).

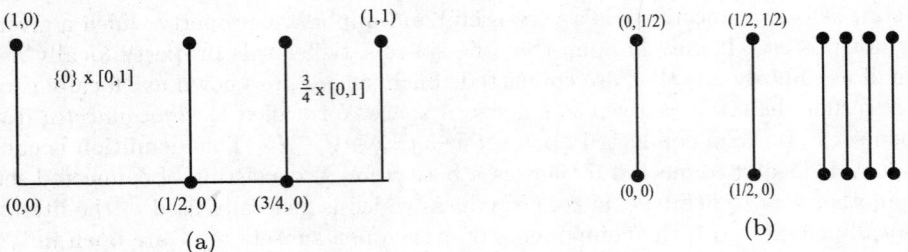

Figure 7.8: (a) The "comb space" is connected but not locally connected. (b) Its subspace U.

But U is precisely $\left(\{0\} \times [0, \frac{1}{2}]\right) \cup \left(A \times [0, \frac{1}{2}]\right)$, which is not locally connected, as can be seen from Figure 7.8(b). (Verify rigorously!).

Here is a useful characterization of local connectedness.

Theorem 7.26. A space X is locally connected iff the components of every open subset of X are open in X.

Proof. Suppose that X is locally connected. Consider a component, C, of any open subset, U, of X and let $x \in C$. By local connectedness of X, we can choose a connected neighborhood V of x s.t. $x \in C \subseteq U$. Since V is connected, it must lie entirely in C because C is a maximal connected subset. Therefore, C is a neighborhood of each of its points and hence C is open in X.

Conversely, suppose that components of open sets in X are open. Let $x \in X$ and let U be any open set containing x. Then, by hypothesis, the component C of U that contains x is open in X and $x \in V \subseteq U$. Hence X is locally connected. □

Application 7.9. Consider $(\mathbb{Q}, \tau_u)$ as a subspace of $(\mathbb{R}, \tau_u)$. The components of $\mathbb{Q}$, are $\{x\}$, $x \in \mathbb{Q}$, which are not open in $\mathbb{Q}$, and hence $\mathbb{Q}$ is not locally connected.

The subspace $\mathbb{Z}$ of integers is locally connected because the components of $\mathbb{Z}$ are $\{x\}$, $x \in \mathbb{Z}$, which are open in $\mathbb{Z}$. Recall that $\mathbb{Z}$ has a discrete topology.

Application 7.10. Any non-empty infinite set X with the co-finite topology is locally connected since the component of any open subset U of X is only U, which is obviously open.

Application 7.11. Consider $\mathbb{R}$ with the p–inclusion topology, and let A be any open subset of $\mathbb{R}$. Since $p \in A$, then, as seen before, A has the p–inclusion topology and therefore it is connected. Hence the component of A is A itself. Therefore, $\mathbb{R}$ is locally connected.

Theorem 7.27. Every open subset A of a locally connected space X is itself locally connected.

Proof. Consider an open subset A of X. Let U be any open subset of the subspace A and C any component of U. Since A is open in X, U is also open in X. By Theorem 7.26, we conclude that C is open in X and hence it is also open in A. Thus every component, C, of each open set in A is open and therefore A is locally connected by virtue of Theorem 7.26. □

Here is another simple characterization of a locally connected space in terms of the basic open sets.

Theorem 7.28. A topological space X is locally connected iff there exists a base β for X consisting of connected sets.

Proof. Let $\beta = \{U_i\}_{i \in I}$ be any base for a locally connected space X. Then by the preceding theorem each component C_{i_α} of the open set U_i is open in X. Therefore, the collection β^* consisting of all components C_{i_α} of all the basic open sets U_i is the desired base for X.

Conversely, let β be a base for X consisting of open and connected subsets of X and let U be an arbitrary open set containing any point $x_0 \in X$. Since β is a base for X, $\exists$ an element $V \in \beta$ s.t. $x_0 \in V \subseteq U$. Thus, by definition, X is locally connected. $\square$

Next we show that local connectedness is invariant under the homeomorphism.

Theorem 7.29. Local connectedness is a topological property.

Proof. Let $f : X \to Y$ be a homeomorphism of a locally connected space X onto an arbitrary topological space Y. We need to prove that $f(X) = Y$ is also locally connected. Let $y \in Y$ and let B be any open set in Y containing the point y. Then $\exists\ x \in X$ s.t. $f(x) = y$. Further, $A = f^{-1}(B)$ is open in X containing the point x. Since X is locally connected, $\exists$ an open and connected subset U of X s.t. $x \in U \subseteq A$. Then $f(U)$ is connected (being the continuous image of a connected set). Also, $f(U)$ is open in Y, since f is an open mapping, and $y \in f(U) \subseteq B$. This proves that Y is locally connected. $\square$

Theorem 7.30. Let $\{X_i\}$ be a finite family of locally connected spaces, $i = 1, 2, \cdots, n$. Then the product space $X = \prod_{i=1}^{n} X_i$ is also locally connected.

Proof. Let β_i denote a base for X_i consisting of connected open subsets of X_i for each $i = 1, 2, \cdots, n$. Then it can be verified easily that the collection

$$\beta = \{U_1 \times U_2 \times \cdots \times U_n | U_i \in \beta_i,\ i = 1, 2, \cdots, n\}, \tag{7.34}$$

is a connected base for X. Hence, by virtue of Theorem 7.28, X is locally connected. $\square$

Corollary 7.30.1.

(i) $\mathbb{R}^n$ is locally connected.

(ii) The n−torus $T^n = \mathbb{S}^1 \times \cdots \times \mathbb{S}^1$ is locally connected.

(iii) The n−cube $\prod_{i=1}^{n}[a_i, b_i]$ is locally connected.

The concept of local connectedness is of direct relevance in Geometry when dealing with continuous curves. In an abstract Hausdorff space, M, a continuous curve is a Hausdorff path. (More precisely, by a continuous curve we mean a Hausdorff space M and a continuous mapping of $[0, 1]$ onto M.) For our present purpose we will regard *this* as the entire space X. (In geometrical applications M would be taken to be a Hausdorff space which is separable and connected (at least arc-wise) and has homeomorphisms from its open cover into $\mathbb{R}^n$.)

Theorem 7.31. Every continuous curve is locally connected.

Proof. Let f be the continuous mapping of $[0,1]$ onto the Hausdorff space X. We must show that X is locally connected. Let C be a component of an open subspace U of X. By Theorem 7.26, we have to prove that C is open in X. Let A be a component of $f^{-1}(U)$. Since A is connected, $f(A)$ is connected in U. Since C is a component of U, we see that $f(A)$ is either disjoint from C or is contained in C by the Corollary 7.18.1. It follows from this that $f^{-1}(C)$ is a union of components of $f^{-1}(U)$. Since $f^{-1}(U)$ is open and $[0,1]$ is locally connected, by Theorem 7.26 the components of $f^{-1}(U)$ are open. Therefore, $f^{-1}(C)$ is open and $\left[f^{-1}(C)\right]' = f^{-1}(C')$ is closed. We conclude the proof by observing that since the mapping f is closed, the set $(f \circ f^{-1})(C') = C'$ is closed, thus C is open in X. Hence X is locally connected. □

Example 7. This example is required for Application to Digital Image Processing in Chapter 8. Here we will call $\mathbb{Z}$ a digital line. The basis element for the digital line is defined as follows:

$$J(n) = \begin{cases} \{n\} & \text{if } n \text{ is odd} \\ \{n-1, n, n+1\} & \text{if } n \text{ is even,} \end{cases} \qquad (7.35)$$

see Figure 7.9. Thus every singleton $\{n\}$ containing an odd integer is an open set. Now consider

Figure 7.9: The digital line. Some of the open sets defined in Eq. (7.35) are shown.

some $A \subset \mathbb{Z}$, then one can not find open sets U and V such that $U \cap A \neq \phi, V \cap A \neq \phi$, $U \cap V \cap A = \phi$ but $A \subset U \cup V$. So the digital line is connected. However, suppose there are two distinct points, say 1 and 2. Then there is no disjoint pair of open sets U and V, with $1 \in U$, and $2 \in V$. Thus the digital line lacks the Hausdorff property.

7.10 Exercises

1. Let X be a connected space and $f : X \to Y$ any continuous function. Show that the graph G_f is connected.

2. Prove that $\mathbb{R}^n$ with respect to the Euclidean topology is connected.

3. Show that if A and B are closed subsets of a topological space and if both $A \cup B$ and $A \cap B$ are connected then A and B are also connected.

4. Prove that if A and B are connected subsets of X s.t. $A \cap \overline{B} \neq \phi$, then $A \cup B$ is also connected.

5. Show that any infinite set with the co-finite topology is connected.

6. Show that $\mathbb{R}$ with the open ray topology is connected.

7. Show that the subspace $\mathbb{Q}$ of all rational numbers of the real line $\mathbb{R}$ is totally disconnected. Note that $\mathbb{Q}$ has the discrete topology.

8. Show that a topological space X is connected iff the diagonal $\{(x,x) | x \in X\}$ is a connected subset of $X \times X$.

9. A non-empty subset S of X is said to be *convex* if for each x and y in S, $(1-t)x + ty \in S$ for every $t \in [0,1]$. Show that every convex subset of $\mathbb{R}^2$ is connected.

10. Prove that $\mathbb{S}^1 = \{(x,y) \in \mathbb{R}^2 | x^2 + y^2 = 1\}$ is connected in $\mathbb{R}^2$.

11. Find a connected subset of $\mathbb{R}^2$ which is not convex.

12. Consider the subspace $\mathbb{Q}$ of $\mathbb{R}$. Find all components of $\mathbb{Q}$.

13. Let X be an infinite set endowed with the co-finite topology. Find all the components of X.

14. Show that the Cantor set is totally disconnected.

15. Is $x \sin(1/x)$ path connected? Justify your answer.

16. If $f : X \to Y$ is continuous and X is path connected, is $f(X)$ necessarily path connected? Justify your answer.

17. If $A \subseteq X$ and A is path connected, is $\overline{A}$ necessarily path connected? Verify!

18. If $\{A_\alpha\}_{\alpha \in I}$ is a collection of path connected subsets of X s.t. $\bigcap_{\alpha \in I} A_\alpha \neq \phi$, is $\bigcup_{\alpha \in I} A_\alpha$ necessarily path connected? Give reasons.

19. Let X be any locally connected space and A an arbitrary open subset of X. Is A always a connected subset of X? Verify!

20. Let $f : X \to Y$ be a continuous function of a locally connected space X into a space Y. Is $f(X)$ always locally connected? Verify! (Note that the statement is always true if f is *both* continuous and open.)

21. Is $\mathbb{R}$ path connected in $(\mathbb{R}, \tau_c)$? Verify!

22. Let $X = \mathbb{R}$ with the usual topology τ. Consider
$$\tau^* = \{A \subseteq \mathbb{R} | A = \phi \text{ or } A' \text{ is compact in } (\mathbb{R}, \tau)\}.$$
Show that $(\mathbb{R}, \tau^*)$ is connected.

23. Prove that X is connected iff $\forall\, x, y \in X\, \exists$ a connected subspace A of X containing them. Apply this result to prove that $\mathbb{R}$ with the p–exclusion topology is connected.

24. Show that if $\{A_\alpha\}$ is a non-empty class of connected subspaces of X s.t. $\bigcap_{\alpha \in I} A_\alpha \neq \phi$, then $A = \bigcup_{\alpha \in I} A_\alpha$ is connected. Use this result to show that every Banach space X is connected.

25. Show that if A is a connected subspace of X, then so is $\overline{A}$.

26. Let $X = \mathbb{R}, \tau = \{A \subseteq \mathbb{R} | A' \text{ is finite or } A' \ni p\}$, i.e., τ is the minimal topology generated by the p–exclusion topology together with the co-finite topology.

    (a) Show that $(\mathbb{R}, \tau)$ is compact.

    (b) Verify that every point q of X, except p, is both open and closed, so $\{q\}$ and $X \smallsetminus \{q\}$ separate X. Hence $X = \{q\} \cup (X \smallsetminus \{q\})$ and thus X is disconnected. Now let $p \neq 0$ and let $A = \mathbb{Z}$ or $\mathbb{Q}$, then $A' \ni 0$ and so A is open, although $A = A \cup \{0\}$ is not open.

    (c) Is A a connected subspace of $\mathbb{R}$?

27. Let $X \{x \in \mathbb{N} | x \geq 2\} = \{2, 3, 4, \cdots\}, U_n = \{x \in \mathbb{N} | x \text{ divides } n\}$ for $n \geq 2, \tau$ is generated by U_n. Thus $U_2 = \{2\}, U_3 = \{3\}, U_4 = \{2, 4\}, U_5 = \{5\}, U_6 = \{3, 6\}, U_7 = \{7\}, U_8 = \{2, 4, 8\}, \cdots$. This is called the *divisor topology*. Show that

    (a) X is connected;

    (b) X is not compact.

Chapter 8

Further Directions in Topology and Applications

8.1 Further Directions in Topology

In this book we have tried to introduce the reader to Topology, starting at the very basics quite heuristically and going on to the more formal development of the subject. While the introduction was entirely motivational and provided the historical background, rigour was introduced from the next chapter on. Nevertheless, the attempt was to keep Chapter 2 very simple and avoid much abstract reasoning in the earlier part of Chapter 3. Though the whole of Chapter 3 is 'concrete' in that it is close to geometry (having metrics defined on the space), it was not possible to keep the entire chapter to the more elementary level maintained for most of the book. If you are interested enough in the subject, you might find it worth your while to go back to the latter part of that chapter later, and re-read it. Chapters 4 and 5 were reasonably straightforward, though Chapter 5 must be admitted to be somewhat abstract. The last two chapters are an attempt to introduce you to those aspects of Topology which would be developed in more advanced courses. We have not gone into it in the depth required for an advanced course but these chapters should make such an advanced course more easily accessible.

It is our intention to keep this final chapter as brief as the first chapter was long and as simple and heuristic as that chapter. Here we would like to indicate the directions of application and development of the subject, as opposed to the Introduction where we explained its origin.

As already mentioned, Topology arose from attempts to do Geometry in unusual situations. Primarily, then, Topology goes into developing Geometry. "Is there much left to develop in Geometry?" you ask. Well, yes. The point is that we repeatedly (and increasingly) need to use the Geometry of unusual spaces. Let us look at some examples.

We often need to find our way in the space of solutions of differential equations. This space has higher cardinality than the continuum. Attempts to use Calculus in this 'larger' space are beset with problems arising from the breakdown of our intuition based on the continuum. While the space is infinite dimensional we need to extend usual Calculus to work in such spaces. Questions of compactness and connectedness are vital there. This is, then, part of the major thrust of the recent (and not so recent) developments of Topology.

Of particular relevance are the differential equations that arise in finding the optimal function for some purpose. These equations, called the Euler-Lagrange equations, arise from the use of the Calculus of Variations. More specifically, they arise from varying the function over all *permissible* functions. Now a function is, essentially a path (or class of paths). The question certainly arises whether all the paths under consideration are topologically equivalent. If they are not, the procedure may not remain valid. As such we need to know when the paths can be guaranteed to be equivalent. On the real line, $\mathbb{R}$, all paths between two points are trivially equivalent as there is a unique path between them (see Figure 8.1). This is not the case on the circle $\mathbb{S}^1$, where one can go from N to S by either the right or the left path (see Figure 8.2), which are inequivalent.

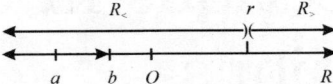

Figure 8.1: Removing the point r from the real line, $\mathbb{R}$, breaks it into two disconnected parts: one to the left, $\mathbb{R}_<$; and the other to the right, $\mathbb{R}_>$. Also, there is a unique path from a to b.

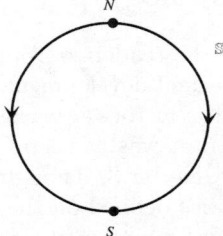

Figure 8.2: There are two paths from N to S, one to the left and one to the right. Further, removing N from $\mathbb{S}^1$ leaves a space that is homeomorphic to $\mathbb{R}$ by a stereographic projection, as shown earlier in Chapter 5 (also, see Figure 1.18).

On a plane, $\mathbb{R}^2$, all paths are again equivalent (see Figure 8.3) as one path between two points can be continuously deformed to any other. This is equally true on $\mathbb{S}^2$ (see Figure 8.4).

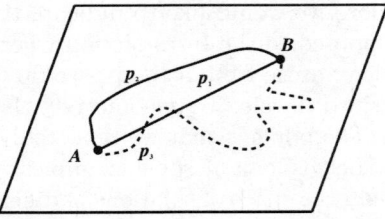

Figure 8.3: Think of a rubber band, p_1, pulled tight about two pins A and B fixed on a sheet. The band could be pulled into the shape p_2 along the dashed line or p_3 along the dotted line without breaking it. This means that there is a homeomorphism transforming p_1 to p_2 or p_3, i.e., $p_1 \cong p_2 \cong p_3$, or all paths are topologically equivalent.

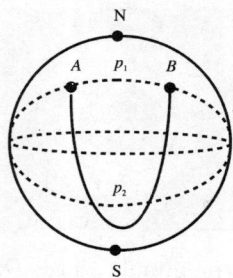

Figure 8.4: Imagine the rubber band now about two pins A and B on a tennis ball. We can pull the rubber band further and further till it goes right round the other side so that $p_1 \cong p_2$. Thus, once again, all paths are topologically equivalent, despite the fact that at first sight p_1 and p_2 look totally distinct.

However, it is not true on the surface of a cylinder, $\mathbb{S}^1 \times \mathbb{R}$, where there are inequivalent paths. In fact, as can be seen from Figure 8.5 there are infinitely many inequivalent paths. One can go from A to B directly, or by going once around the right, or once around the left, or twice around the right, or twice around the left, etc. Again a torus has infinitely many inequivalent paths. Such spaces are said to be *multiply connected*. A connected domain $\mathcal{D} \subseteq \mathbb{R}^2$, which is itself 2-dimensional, is simply connected iff its boundary, $\partial \mathcal{D}$, is connected. If $\partial \mathcal{D}$ consists of n disconnected pieces it is n-times connected. Thus it is doubly connected if the boundary consists of two pieces. For example an annular domain is doubly connected, see Figure 8.6.

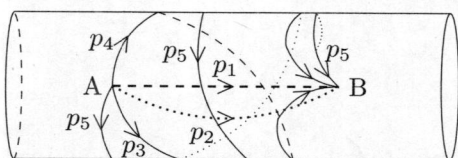

Figure 8.5: Whereas the dotted path $(p_2) \cong$ the dashed path (p_1). These are inequivalent to the path p_3 winding once round the cylinder to the right, p_4 winding once round to the left, p_5 winding twice round to the right, etc. In fact $p_2 \ncong p_3 \ncong p_4 \ncong p_5 \ncong p_2 \ncong p_4, p_3 \ncong p_5$. There are infinitely many such inequivalent paths.

A plane with a hole punched out of it, $\mathbb{R}^2 \setminus \{O\}$, is multiply connected. As may be seen from Figure 8.7, this space is topologically equivalent to a cylinder, since there are homeomorphisms from one space into the other. Such identifications of spaces play a vital role in the use of Topology for applications. We will not pursue this matter further, here, but will turn our attention to another aspect of multiply connected spaces.

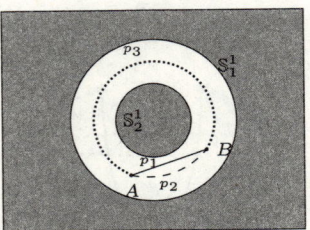

Figure 8.6: The path from A to B in the annular disc, D; p_1, p_2, p_3; are such that $p_1 \cong p_2 \not\cong p_3$. The boundary, ∂D consists of two circles $\mathbb{S}_1^1$ and $\mathbb{S}_2^1$, i.e., $\partial D = \mathbb{S}_1^1 \cup \mathbb{S}_2^1$ and $\overline{\mathbb{S}_1^1} \cap \overline{\mathbb{S}_2^1} = \phi$. Thus ∂D consists of two disconnected pieces.

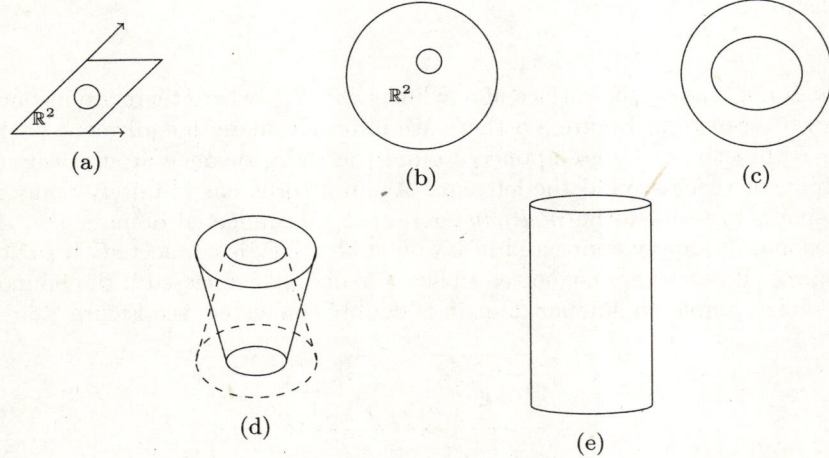

Figure 8.7: (a) Imagine an infinite rubber sheet with a hole in it. (b) We can think of it in polar instead of Cartesian coordinates, with the origin at the hole, O. (c) We can now "open up" the origin to the circle $\bar{r} = 1$ and "pull in infinity" to the circle $\bar{r} = 2$ by the transformation $\bar{r} = (2r+1)/(r+1)$. (d) Now imagine this annular rubber sheet pulled out over a unit height so that the smaller circle is below and the larger one on top. (e) Now straighten out this "cut off" cone into a cylinder of average radius. This corresponds to using the polar coordinates $(\bar{r}, \theta)$ as the cylindrical coordinates, $(\theta, \bar{z})$, with $\bar{z} = \bar{r}$. Here the cylinder goes from $\bar{z} = 1$ to $\bar{z} = 2$. Now we can stretch the cylinder to infinity by the transformation $z = \tan\{(\bar{z} - \frac{3}{2})\pi\} = \tan\left(\frac{r-1}{r+1} \cdot \frac{\pi}{2}\right)$. Clearly $r = 0$ now corresponds to $z = -\infty$ and $r = \infty$ to $z = +\infty$. This also gives the transformation from $\mathbb{C} \setminus \{0\}$ to the cylinder. Here we replace (x, y) by $x + \iota y$ or, for a complex number u, $r = |u|$ and $\theta = \tan^{-1}(Im(u)/Re(u))$.

If we cut the line at one point it breaks into two disconnected pieces, see Figure 8.1. This may be seen as

$$\left.\begin{array}{r} \mathbb{R} \setminus \{r\} = \mathbb{R}_{<r} \cup \mathbb{R}_{>r} \ s.t. \\ \overline{\mathbb{R}_{<r}} \cap \mathbb{R}_{>r} = \mathbb{R}_{<r} \cap \overline{\mathbb{R}_{>r}} = \phi. \end{array}\right\} \quad (8.1)$$

However, if we cut a circle we get one connected piece, see Figure 8.2. This may be seen as

$\mathbb{S}^1 \setminus \{N\} \cong \mathbb{R}$. Consider, now, the next dimensional generalization. Cutting the plane into two pieces is achieved by subtracting a line from it, i.e., $\mathbb{R}^2 \setminus L$ where $L = \{(x, y) | y = mx + c\}$ gives two disconnected pieces: $\mathbb{R}^2_<; \mathbb{R}^2_>$; given by

$$\left. \begin{array}{c} \mathbb{R}^2_> = \{(x,y) | y > mx + c\}, \mathbb{R}^2_< = \{(x,y) | y < mx + c\}, \\ \overline{\mathbb{R}^2_>} \cap \mathbb{R}^2_< = \mathbb{R}^2_> \cap \overline{\mathbb{R}^2_<} = \phi. \end{array} \right\} \quad (8.2)$$

This is depicted in Figure 8.8. Similarly, cutting the surface of a sphere by a circle breaks it into two disjoint, disconnected pieces, i.e., $\mathbb{S}^2 \setminus \mathbb{S}^1$ where $\mathbb{S}^1 = \{(x, y, 0) | x^2 + y^2 = 1\}$, $\mathbb{S}^2$ being the surface of a unit sphere, we get the upper and lower heimispheres:

$$\left. \begin{array}{c} \mathbb{S}^2_> = \{(x, y, z) | x^2 + y^2 + z^2 = 1, z > 0\}, \\ \mathbb{S}^2_< = \{(x, y, z) | x^2 + y^2 + z^2 = 1, z < 0\}, \overline{\mathbb{S}^2_>} \cap \mathbb{S}^2_> = \mathbb{S}^2_< \cap \overline{\mathbb{S}^2_>} = \phi. \end{array} \right\} \quad (8.3)$$

This is depicted in Figure 8.9. However, cutting a cylinder we may get two pieces or one, see Figure 8.10.

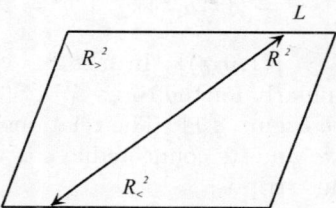

Figure 8.8: Cutting the plane, $\mathbb{R}^2$, by a line L amounts to subtracting L from $\mathbb{R}^2$ leaving $\mathbb{R}^2_<$ below and $\mathbb{R}^2_>$ above L. Clearly these are two disconnected pieces. Thus the plane is simply connected.

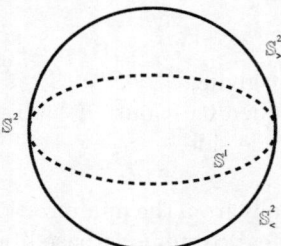

Figure 8.9: Cutting the surface of the sphere $\mathbb{S}^2 = \{(x, y, z) | x^2 + y^2 + z^2 = 1\}$ by removing the $\mathbb{S}^1 = \{(x, y, 0) | x^2 + y^2 = 1\}$, we get two hemispheres, $\mathbb{S}^2_<$ and $\mathbb{S}^2_>$ which are obviously disconnected. Thus the surface of the sphere is disconnected.

$$\mathbb{A} = \mathbb{S}^1 \times \mathbb{R} \setminus \mathbb{S}^1 = \mathbb{A}_< \cup \mathbb{A}_>, \quad (8.4)$$

consists of two disconnected pieces but

$$\mathbb{B} = \mathbb{S}^1 \times \mathbb{R} \setminus L = \mathbb{B}_- \cup \mathbb{B}_+, \quad (8.5)$$

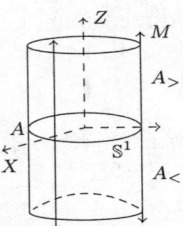

Figure 8.10: Cutting the cylinder $\mathbb{S}^1 \times \mathbb{R} = \{(x,y,z) \,|\, x^2+y^2+z^2=1\} = \mathbb{A}$ by removing the circle $\mathbb{S}^1 = \{(x,y,0) \,|\, x^2+y^2=1\}$ breaks it into two disconnected pieces, but cutting it by removing the line $L = \{(1,0,z)\}$ leaves it in one sheet. To cut into two pieces this way we must remove another line, $M = \{(0,1,z)\}$, say. We then get the front part (positive quadrant sheet) and the rest.

consists of one, where the cylinder may be written as

$$\mathbb{S}^1 \times \mathbb{R} = \{(x,y,z) \,|\, x^2+y^2=1\}, \tag{8.6}$$

$\mathbb{S}^1 = \{(x,y,0) \,|\, x^2+y^2=1\}$ and $L = \{(1,0,z)\}$. In fact $\mathbb{B}_- \cong \mathbb{R}^2$. To break it into two pieces another line must be removed. Similarly for the torus $\mathbb{S}^1 \times \mathbb{S}^1$, if we cut it by removing a circle, $\mathbb{S}^1 \times \mathbb{S}^1 \setminus \mathbb{S}^1$, we get a cylinder see Figure 8.11. The relationship between the number of "cuts" required to split the space into two and its connectedness is worth pondering. Again, consider what the connectedness of a Möbius strip is.

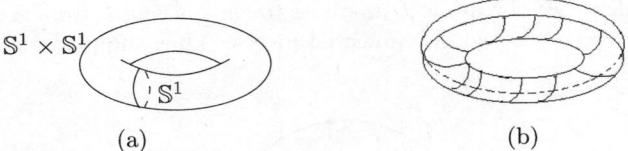

(a)   (b)

Figure 8.11: If we make a vertical cut giving $\mathbb{S}^1 \times \mathbb{S}^1 \setminus \mathbb{S}^1$ as shown in (a), we get a "tube" which can be deformed into a cylinder by opening it out. A lateral cut, as shown in (b), again gives a cylinder by "unfurling" the larger circle side.

Since all Dynamics derives essentially from the minimization of a function called the Lagrangian, the study of Dynamics and Dynamical Systems is based on the study of the connectedness of the space of 'permissible functions'. This applies not only for Classical Mechanics of Continuous Media but for its more modern version of Classical Field Theory and its further development of Quantum Field Theory. Nowadays Topology has become crucial for these studies, and their sub-branches: Conformal Field Theory; and Superstring Theory (see for instance [Nash, 1997]). In these there is great physical relevance of a topological invariant of a path, called the *winding number* [Ueno, 2003]. This number can be understood in the context of the example of a plane with a hole punched in it, see Figure 8.12. The plane is punctured at P and any path from Q to R has to bypass P. The path γ_1 goes in the clockwise direction, but does not wind around P, so the winding number for γ_1 is 0. Similarly for the the path γ_2, which goes in counterclockwise direction, the winding number is 0. But the path γ_3 winds around P once, thus the winding number for γ_3 is

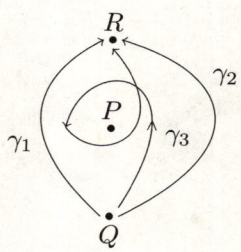

Figure 8.12: For a plane punctured at P, we consider paths from point Q to another point at R. Clearly the path γ_1 going to the left of P cannot be continuously deformed to the path γ_2, going to the right of P. Further, γ_3 is obviously topologically distinct from γ_1 and γ_2 as it has wound once around P. Paths could wind any number of times about P, which would all be distinct.

1. None of these paths can be interconverted by a homeomorphism, i.e., $\nexists\, f\ s.t.\ f : \gamma_i \to \gamma_j (i \neq j)$ and f is a homeomorphism.

The very existence of limits, derivatives, integrals, etc., becomes doubtful when dealing with the calculus of functions. This rigorous study is called Functional Analysis. Currently it is extensively applied in the study of Dynamical Systems. In general, the differential equations derived from here are non-linear. The solutions of non-linear differential equations can, in general, vary wildly. Sometimes they lead to very specific repetitive orbits, but some times they can cover the entire space (or a higher dimensional subspace). When the latter situation arises it is said to be in the 'chaotic region'. While the study is actually of the *stability* of the solution, or the *branching* of the solutions, it is popularly called the study of Chaos. This term is misleading but has captured the public imagination.

In special cases the chaotic regions are *not* the entire space but give patterns which repeat themselves (some times with minor variations) endlessly. One famous example of this is the Lorenz 'strange attractor'. In this (see Figure 8.13) two sources first 'attract' and then 'repel' a 'test particle'. Another famous example is the Mandlebrot set (see Figure 8.14).

These give geometries that, in some sense, lie between two dimensions and have been called 'fractional dimensional' or *fractal* geometries. The Cantor set can be regarded as a fractal of dimension $0.631 \cdots$. Again the *Koch curve*, shown in Figure 1.13, can be regarded as having dimension $1.2618 \cdots$. The definition of the fractal dimension used to obtain these values is

$$d = \lim_{\epsilon \to 0} \frac{\log N(\epsilon)}{\log(1/\epsilon)}, \qquad (8.7)$$

where $N(\epsilon)$ is the number of pieces, each of length $1/\epsilon$ into which the figure is broken. In the limit the length of the perimeter of the figure may become infinite. This investigation also lies in the domain of Topology [Edgar, 2007].

From the viewpoint of Mathematics itself, as opposed to the Mathematics developed for application, Topology has occupied the central position in many fields. Topology pervades subjects like Knot theory (see the chapter on Knot theory in [Goodman, 2005] for a readable introduction. For further details and applications see [Adams, 1994]). Subjects which might have been regarded as a part of Algebra, through the development of Algebraic Topology, have now begun to rely heavily on topological reasoning. Thus, for example, Homology Theory and

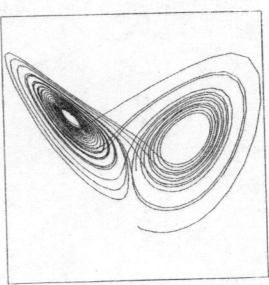

Figure 8.13: The Lorenz "strange attractor" is the set of three equations which generate solutions that can be traced out in two dimensions (suppressing a third): $\dot{x} = \sigma(y-x), \dot{y} = \sigma x - y - xz, \dot{z} = -\beta z + xy$, where σ, ρ, β are given constants. It has various critical points, e.g. at $\sigma = 10, \beta = 8/3, \rho = 28$. The "strangeness" comes for larger ρ. Regarding the trace as giving the orbit of a particle, it looks as if the particle is being attracted by two sources, but as it comes close enough to each source it is repelled again. (A similar behaviour can arise with one charged massive point gravitational source in General Relativity.)

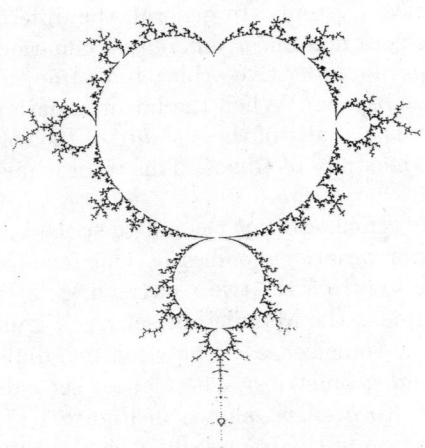

Figure 8.14: The Mandlebrot set is generated by iterating the transformation $z \to z^2 + c$. It generates an extremely complicated set which is nearly (but not quite) self-similar at different scales. Self-similarity is the true hallmark of fractals.

particularly Cohomology need Topology as an essential ingredient, see for instance [Hatcher, 2002]. The theory of Measures and Integration, Functional Analysis and of course Geometry cannot be done justice to without Topology.

Till recently Topology was one of the last bastions of 'pure' Mathematics. Now Topology is becoming a *must* for theoretical physicists and can be expected to rapidly extend its domain of influence. Again, Topology is required in Economics and is becoming important in Game Theory and Decision Making. Part of the reason is its use in the Bolzano-Weierstrass theorem, which is used for finding optimal solutions, or proving that there is no unique optimal solution available. However, it is also needed to deal with situations for which it was taken for granted that there Mathematics cannot be applied but, with different topologies it may be that it could. The time has come when a mathematician who does not know Topology will be regarded as half-baked. This book is an attempt to make the subject accessible to all who need it. It provides the fundamental language and the type of reasoning involved in Topology. For further information on the subject more advanced books are needed, see for instance [Crossley, 2006].

8.2 Application of Multiply Connected Spaces to Wormholes

For fans of science-fiction, such as *Star Trek*, one of the most interesting applications of topology is to "wormholes". Numerous science-fiction films and books have spaceships taking "shortcuts" through wormholes so that they can break the speed of light limit. This name was given to match "black holes", by the person who invented the name of the black hole, John Wheeler [Ruffini and Wheeler, 1971]. A black hole is a region of space where the gravity is so strong that even light cannot escape, and so it is totally black, but things can fall into it, so it is like a hole. The way this is explained is that gravity causes the space around it to curve, something like putting a steel ball on a rubber sheet. If the ball is heavy enough, the sheet will acquire a dip and if it is very heavy it will sag a lot, see Figure. 8.15. A black hole is like a ball that is dense

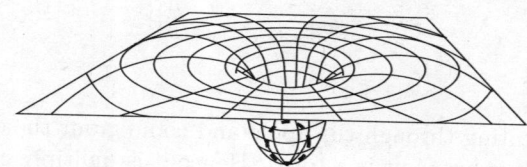

Figure 8.15: Rubber-sheet geometry. A ball is put on a rubber-sheet. If we replace a larger ball by a smaller ball of the same mass, the sheet will sag more around the ball. In this picture we see the curve of the rubber sheet with the ball shown as a dashed line, to indicate that it has been removed but the sheet is curved as if it was there.

enough to curve the sheet to this radius. Think of the original sheet having grid lines on it so that it will bend down till it gets to that radius and will have nowhere to continue. However, as we know from Calculus, curves cannot just disappear, but must have a definite end-point or go on forever. To avoid this problem, Einstein and Rosen added an extension by reflecting the sheet where it comes closest to the gravitational source, see Figure. 8.16, so that it continues forever. This is called the Einstein-Rosen bridge. Now imagine the ball getting still smaller and denser, so that the sheet goes into the black hole and then comes back out. Finally, imagine a point particle, in which case the sheet comes to a halt at a vertex.

The rest of the story for the name comes from the apocryphal story of Newton getting the idea of his law of universal gravitation by being hit on the head by an apple. (One wonders if it is the same apple that Adam and Eve ate and acquired the knowledge that led to their being expelled from Eden.) This led Misner, Thorne and Wheeler to put a large apple on the cover of their book *Gravitation* [Misner et al., 1973]. Now imagine one side of the Einstein-Rosen bridge starting at one point of the black hole and a worm entering there and exiting where the other end of the bridge lies on the apple, see. Figure. 8.17(a).

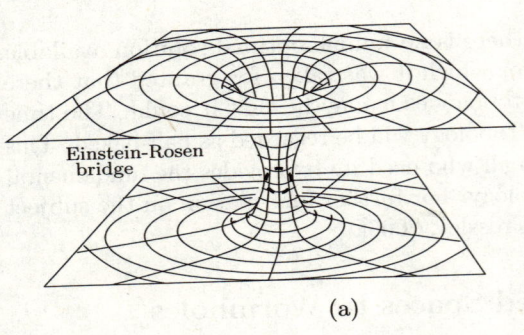 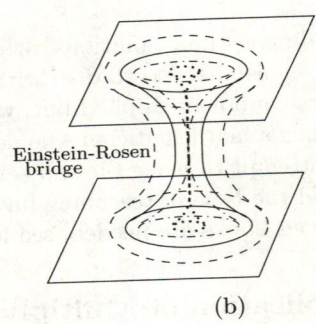

Figure 8.16: Imagine the size of the ball (thick, dashed) being successively reduced to the radius of the black hole and then smaller and smaller radii. In (a) the ball has just broken through the one sheet to the extension of the spacetime. In (b) as the ball gets still denser the "throat" narrows still more, till in the limit it gets "pinched off" to a vertex and restarts at a vertex lower down.

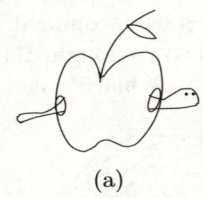 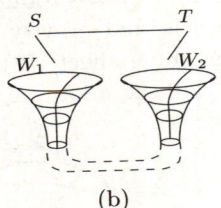

Figure 8.17: (a) A worm eating through the apple and coming out the other end. Now the space is no longer topologically a sphere but is a torus. Hence it is multiply connected. The hole takes no time to traverse but going around the apple does, so the wormhole provides a shortcut to go from one side to the other. (b) The apple is "straightened out", so that the hole gets bent into a U. Label the point where the worm entered the apple as W_1 and the point where it exited as W_2. Now before and above W_1 is a star, S, and beyond and above W_2 is an observer T. Light going directly along ST will take longer to reach than light that enters at W_1 and exits at W_2 if $SW_1 + W_2T < ST$. This makes it into a shortcut.

You now see why it is a *worm*hole? There is one direct route through the apple and a topologically distinct one along the surface of the apple, so the space is multiply connected. This route happens to take no time to traverse, while the other takes the normal time. *This* is why it is a shortcut – you just fall through it to the other side instantaneously. Now take it to be a rubber apple and imagine straightening out the surface, so that the wormhole looks deformed into a flat U-shape, see Figure. 8.17(b).

In this case the surface represents the usual Universe and the wormhole provides a quick way from point W_1 to point W_2 in the Figure. If there is a star, S, over W_1 and its light is seen by someone at T, it can either go along ST, or SW_1W_2T. If $SW_1 + W_2T < ST$, we see that the light from the star will be seen as coming from two different directions and coming later along the usual path than through the wormhole. If we can identify that it is the same star we will have "seen the wormhole". This has been suggested as a test for whether there are actual wormholes in the Universe.

8.3 Application of Winding Numbers to Magnetic Fields and Monopoles

Everybody has seen bar magnets with a North and a South pole, called a dipole, and all are familiar with charged particles like electrons and protons. We know that hydrogen atoms behave like electric dipoles (as they have a positive and a negative charge) but we have never seen a magnetic charge, or monopole. Why is that? A simple answer would be that they do not exist in Nature. If we try to break a bar magnet into just a South pole or just a North pole, all we succeed in doing is to break the big bar into two smaller bars. In other words we break the one big dipole into two smaller dipoles so, if it does not exist in Nature maybe it cannot be artificially produced. The theory for electricity and magnetism has remarkable similarities between electricity and magnetism, but also some big differences, which reflect the absence of magnetic monopoles. The field of a dipole shows "handedness" (which the physicists call *chirality*), in that negative charges spiraling from the South to the North pole go along a right-handed spiral, while positive charges go along a left handed spiral. A neutral particle goes straight instead, see Figure 8.18. For example, if the negative charge is an electron and the positive a proton, since the proton is much more massive (1836 times), it will be deflected by the magnetic field much less (1836 times) than the electron. As such, if the proton has a winding number -1, the electron will have one of $+1836$.

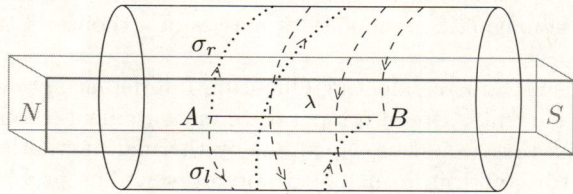

Figure 8.18: A cylinder surrounding a bar magnet has points A and B marked on it. A neutral particle travels directly from A to B along a (hard) straight line λ; a negatively charged particle goes along a (dotted) right-handed spiral, σ_r, shown going around three times between the same two points; and a positively charged particle going along a (dashed) left-handed path, σ_l, twice. Clearly the three types of paths are topologically different. The winding number of λ is 0, of σ_r is $+3$ and of σ_l is -2.

The Universe has an all-pervading background radiation in the microwave band, corresponding to a temperature of $2.7255°K$. Such radiation was predicted by George Gamow if the Universe started with a Big Bang [Gamow, 1946]. From the observed temperature, it was possible to work out how much matter of the usual type (protons, electrons, and neutrons) there must be in the Universe. The problem was that the prediction worked *too well*. Why was the temperature so uniform? The matter would not have had time to reach the same temperature everywhere that we could now see, to that accuracy (one in a hundred thousand). For this purpose an extension of the unified theory of Electromagnetic and Weak Nuclear forces of Sidney Glashow [Glashow, 1980], Abdus Salam [Salam, 1980] and Steven Weinberg [Weinberg, 1980] was invoked. The standard theory has a Higgs particle with a mass of 125 times that of the proton, which is responsible for all particles acquiring masses. In the extension the Strong Nuclear Force was to be unified with the other two and the corresponding Higgs particle would be more than a trillion times as much.

When the Universe cooled to the energy corresponding to that mass ($E = mc^2$), it would undergo exponential expansion, much like economic inflation raises prices. Thus, Alan Guth, who proposed this model, called it "cosmic inflation" [Guth, 1981]. Because of the sudden expansion, regions that had earlier been close would appear to have always been too far away to achieve thermal equilibrium.

While inflation solved the problem, it raised its own new problems, which are what we are interested in. Sudden expansion would lead to sudden cooling. Like cooling water produces ice, the state of the Universe would undergo a qualitative change, called a phase transition. When the cooling is sudden, ice crystals form in the water at different places at the same time, and where they meet, there are defects. (This is how defects in diamonds are also produced, by pressure applied too suddenly.) In the case of ice, the defects occur along the line where ice crystals meet, and so they are two-dimensional, surface defects, called grain boundaries. In the Universe as a whole we need to consider a four-dimensional spacetime, that can have three, two, one, or zero-dimensional topological defects. The first are called "textures" and are unstable, and would unravel over time. The next are here called "domain walls", and would cause distortions in the background radiation that are not seen. The one-dimensional defects would be "cosmic strings" and could have caused the large scale structures that we see in the Universe. The zero-dimensional defects would be "point-like", but would actually be bubbles of the earlier phase of the Universe trapped in the new phase. In that phase there would have been total symmetry between electricity and magnetism, so that these would be — magnetic monopoles! There was a tremendous surge of studies searching for these magnetic monopoles of masses of a thousand trillion times that of the proton.

Physicists had been using Superconducting Quantum Interference Devices (SQUIDs) to detect extremely small electric currents. Room temperature measurements would not be good enough for such small currents, as they would be wiped out by thermal fluctuations. However, a current set up in a superconductor could, in principle, go on forever. The problem would have been to now see that the current has been set up. This was done by quantum interference. The SQUID consists of a loop (or loops) of a superconducting wire, connected to a quantum interference device that "reads" the signal and amplifies it, and records it on a rotating drum, see Figure 8.19(a) and Figure 8.19(b). Blas Cabrera tried using a SQUID of about 25 cm^2 to detect the monopole [Cabrera, 1982]. The idea was that if a (magnetic) monopole passed through the loop it would set up a current in the SQUID and it would show a jump up, if it was a North pole, or a jump down, if it was a South pole. *If one came, we would see a topological defect!* How would we know that it was not a spurious signal? If a dipole passed through, one pole would set up the current and as the other passed through, the second would switch it off, so there would be no signal, if the time of passage were too short, and a blip if there was enough time. If there were a current surge for some other reason there would be the step up or down, but there would be fluctuations about it. However, if the monopole passed through it would leave a perfect step function as a trace on the drum (Figure 8.19(c)). Within a month Cabrera claimed to have the signal recorded. However, when other people tried to use much larger SQUIDs for much longer, they got no signal, so the claim was not generally believed. A theoretical problem arose that the number of such massive monopoles would cause too much matter in the Universe to be possible and the monopoles would "switch the Sun off". Andre Linde presented a new inflationary theory [Linde, 1982] that reduced the rate at which the change of phase occurs allowing much more inflation with fewer defects, so that only one would be found in the entire observable Universe. Making fun of Cabrera, people started saying, "There is only one magnetic monopole in the Universe — *and Cabrera saw it!*"

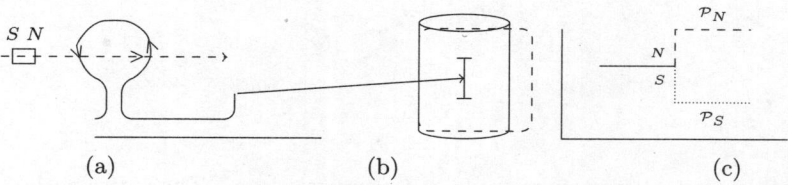

Figure 8.19: A SQUID shown in (a) is connected to a pen resting on a slowly rotating drum shown in (b). As a monopole goes through the SQUID, the trace jumps up, or down, according as it is a North or South pole. The path traced for North (dashed) and South (dotted) pole is shown in (c), and is denoted by $\mathcal{P}_N$ and $\mathcal{P}_S$, respectively.

8.4 Application of Connected Sets to Digital Image Processing (DIP)

In Example 7 of Chapter 7, we discussed the Digital line $\mathbb{Z}$. In set terms a digital picture is a two-dimensional array, i.e., a subset of $\mathbb{Z}^2$. A *pixel* is a single element of the digital array, say P, given by $(m,n) \in \mathbb{Z}^2$, see Figure 8.20. Recall in Section 1.3 we defined that in the four-colour problem, adjacent means that some of the sides are being shared. Similarly here two points are adjacent if they share the same m or n, but the other number is different from the value for P by one either side. On the other hand, one could even call a point that is diagonally removed from P by one up or down on both numbers as shown in Figure 8.20. The points m, n $(m \neq n)$ are called α−adjacent (or 4−adjacent) when $(m,n) \in \alpha$ iff $|m_i - n_i| \leq 1$ for $i = 1, 2$. They are called ω−adjacent or (8−adjacent) if $(m,n) \in \omega$ iff $\sum_{i=1}^{2} |m_i - n_i| = 1$.

The neighborhood of P can be defined by P or the set of points that are adjacent to it. If they share a side, we say it is a $4 - nbd$ and if they even only share a point, apart from those sharing a side, we call it an $8 - nbd$.

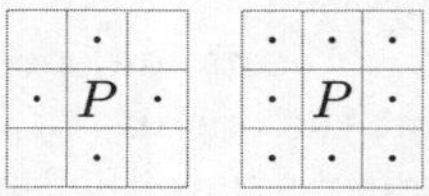

Figure 8.20: $4 - nbd$ and $8 - nbd$ in a Digital space. The former only includes the top, bottom, left and right neighbors. The latter includes these as well as top right, top left, bottom right and bottom left neighbors.

The notion of connectedness is essentially the same as that in Section 7.6. An ordered sequence of points $m_i, \ldots, m_k$, in a digital space is called a *path* such that m_i is adjacent to m_{i-1}, for $i \in \{1, \ldots, k\}$. The points m and n are said to be *connected* if there exists a path between them. Two pixels P and Q are said to be 4-connected (8-connected), if the path joining them contains only 4-adjacent (8-adjacent) pixels, see Figure 8.21 [Herman, 1998]. The boundary between two subsets A and B of $\mathbb{Z}^2$ is defined to be

$$\partial(A, B) = \{(m, n) \ s.t \ m \in A, n \in B, (m, n) \in \omega\}, \tag{8.8}$$

see Section 2.3. Let m_0 be starting point of a path in the digital space and m_n, be the end point.

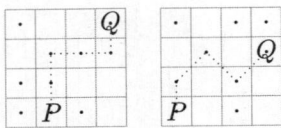

Figure 8.21: The figure on the left is an example of 4−connectedness. The entire path is connected using only $4 - nbds$. The right figure is an example of 8−connectedness, as the path is connected using $8 - nbds$.

Then it is called a closed curve if $m_0 = m_n$. Furthermore if the path does not self-intersect, it is said to be a simple closed curve. As such, any curve in $\mathbb{R}^2$ with integer vertices is a curve in the digital space $\mathbb{Z}^2$. Thus we extend the standard Jordan Curve Theorem for digital spaces as follows:

Theorem 8.1. Given a digital simple closed curve $\mathcal{C}$, $\mathbb{Z}^2 \setminus \mathcal{C}$ is separated into two disjoint sets. One of which is the interior and the rest is exterior.

Figure 8.22: (a) A black square formed out of individual pixels, (b) and its boundary. This to some extent explains how Youtube and other websites provide us the option to view a video in Low Quality, and still clearly recognize objects and people in the video. The square in (a) is "meaty", because it has more pixels. But the one in (b) is the skeleton of the square in (a). It is the same square inasmuch as both the squares have the same size. Thus if we take out a few pixels from a photo, that would not harm. In a much bigger photo, the harm would reduce furthermore.

In DIP, the boundaries of digital images are of great significance. One application is discussed below [Krantz, 2009]. As a subset of $\mathbb{Z}^2$, consider a digital image that contains a finite set of pixels say S. As defined in Equation (8.8), the boundary of a digital image is the set of points such that they have at least one neighbor that is not in S, see Figure 8.22(b). If one stores the full image, it will use up 64 pixels worth of data as shown in Figure 8.22(a), while storing only the boundary, reduces the pixels to 28 as shown in Figure 8.22(b).

8.5 Application of Graphs to Electronic Chip Design

In Figure 1.3 there are two paths from Island 1 to each of the two shores. And one path from Island 2 to both of the shores. If both Islands and Shores are taken to be mere vertices, and call the seven bridges as edges, then the object that emerges is a *graph*, and is studied in Graph Theory as alluded to in Section 1.3.

A printed circuit board (PCB) is a similar object in that there also is a set of terminals connected by copper wire connections among some of the terminals (Instead of imagining people walk on the bridges, consider electric signals travelling through the wires).

(a)   (b)

Figure 8.23: (a) An arbitrary electronic circuit is shown. The hollow dots represent terminals and the connections among them are shown by solid lines. (b) An arbitrary graph is shown. The possibility of deploying it as a circuit requires topological considerations. From left to right, the path from second last to the last terminal is multiply connected, physically this phenomenon is called short circuit, and thus cannot be used to model an electronic circuit.

Using graphs to model electronic circuits puts topological constraints on the type of graphs that one can use. For instance, a short circuit occurs when two ends of a wire are joined to each other, forming a loop. Thus the graph in Figure 8.23(b), cannot be a model for an electronic circuit.

Besides loops, a graph may also have parallel edges i.e., two vertices being connected by more than one edge. To this end, we define a *simple graph* to be one that has no loops and no parallel edges.

In designing an electronic circuit, it is not necessary that every terminal (vertex) of the circuit (graph) be connected to every other terminal (vertex). As such, we define a *complete bipartite graph* $K_{s,t}$ so that the vertices are partitioned into two classes, say S and T. There is a vertex from each member s of S to each member t of T, but none within the members of each of the two sets. See $K_{2,3}$ in Figure. 8.24.

Figure 8.24: The lower three nodes connect to the upper two nodes. But notice that there is no connection amongst the lower three nodes. Similarly the upper two nodes, though connected to the lower three, are themselves detached.

A PCB is a two-dimensional sheet. The circuit is (preferably) so designed that the connections between the terminals do not intersect each other. We define a *planar graph* to be one that can be imbedded in a two-dimensional plane without any of the edges crossing one another. $K_{2,3}$ is clearly not a planar graph. K_4 shown in Figure. 8.25 is a complete simple graph. Each vertex is connected to every other vertex. If the edge connecting the vertices a and c is modified to that shown, then the intersection present in the left figure below is eliminated. Thus K_4 is a planar graph.

The Euler number is a topological invariant defined as $E = v - e + f$, where v is the number of vertices, e is the number of edges, and f is the number of faces. For any graph with $v \geq 3$, the

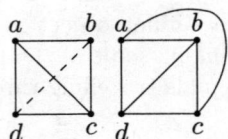

Figure 8.25: The figure on the left is K_4. The dashed diagonal is underneath the solid diagonal. We would need 3-space to keep the solid diagonal from touching the dashed diagonal. However, when we classify K_4 as a planar graph, what we mean is that going to 3-space is not necessary. The figure on the right shows that we can pull the solid diagonal, without disarranging its connection to the nodes, and staying in the 2-space.

following inequality is true.
$$3f \leq 2e. \tag{8.9}$$

We will use Inequality (8.9) along with the Euler number to obtain an upper bound on the number of edges a planar may have. In a graph, a face cannot be formed with fewer than three vertices. So Inequality (8.9) is essentially counting the number of edges, but only face-by-face. Read as $2e \geq 3f$, Inequality (8.9) says that the number of edges is at least three times the number of faces. The factor of two expresses that each edge is shared by two faces.

Thus using Inequality (8.9) and the Euler number, it is a simple exercise to obtain $e \leq 3v - 6$, which is the criterion a graph must fulfil in order to be a planar graph.

Appendix

As promised in Chapter 5, here is a simplified proof of the Urysohn's Metrization Theorem [Simmons, 2004].

Theorem 8.2 (Urysohn's Metrization Theorem). Let $\mathbb{R}^\infty$ denote the infinite-dimensional Euclidean space with l_∞ norm defined on it. If X is a second countable, normal space, then $\exists$ a homeomorphic map $f : X \xrightarrow{into} \mathbb{R}^\infty$, and X is thus metrizable.

Proof. Since X is second countable, there exists basis set $\{G_1, G_2, \ldots\}$. Now by normality of X, there exists G_i s.t. $x \in G_i \subseteq \bar{G}_i \subseteq G_j$. The set of ordered pairs (G_i, G_j) are countably infinite and thus can be arranged in a sequence $P_1, P_2, \ldots$. The subspace of R^∞ that we have chosen is $[0,1]$. By Urysohn's Lemma, $\exists f_n : X \to [0,1]$ for each ordered pair $P_n = (G_i, G_j)$ s.t. $f_n(G_i) = 0$ and $f_n(G'_i) = 1$. For each $x \in X$, define $f(x) = \{f_1(x), \frac{f_2(x)}{2}, \ldots, \frac{f_n(x)}{n}, \ldots\}$.

This mapping is bijective, now we need to show that it is bicontinuous. By virtue of the fact that $\sum_{n=1}^{\infty} \frac{1}{n^2}$ is convergent (convergence of this series is also known as Basel Problem), and by the Cauchy convergence criterion, $\exists n_0$, s.t.

$$||f(y) - f(x_0)||^2 = \sum_{n=1}^{\infty} \left|\frac{f_n(y) - f_n(x_0)}{n}\right|^2 < \sum_{n=1}^{n_0} \left|\frac{f_n(y) - f_n(x_0)}{n}\right|^2 + \frac{\epsilon^2}{2}.$$

By continuity of $f_n, \forall n = 1, \ldots, n_0$, $\exists$ a nbd H_n of x_n s.t. $y \in H_n \implies \left|\frac{f_n(y) - f_n(x_0)}{n}\right|^2 < \frac{\epsilon^2}{2n_0}$. Let $H = \bigcap_{n=1}^{n_0} H_n$. Then H is a nbd of x_0 s.t. $y \in H \implies ||f(y) - f(x_0)||^2 < \epsilon^2$, or $||f(y) - f(x_0)|| < \epsilon$.

Finally, we show that f^{-1} is continuous. Consider $f^{-1} : f(X) \to X$. Recall that G_j is the second member of some ordered pair $P_{n_0} = (G_i, G_j)$ s.t. $x_0 \in G_i \subseteq \bar{G}_i \subseteq G_j$. Now if we choose $\epsilon < \frac{1}{2n_0}$, then notice that $||f(y) - f(x_0)|| < \epsilon \implies \sum_{n=1}^{\infty} \left|\frac{f_n(y) - f_n(x_0)}{n}\right|^2 < \left(\frac{1}{2n_0}\right)^2$. In particular for $n = n_0$, $|f_{n_0}(y) - f_{n_0}(x_0)| < \frac{1}{2}$. Since $x_0 \in G_i$ and $f_n(\bar{G}_i) = 0 \ \forall \ n$, $f_{n_0}(x_0) = 0$. Thus $|f_{n_0}(y) - f_{n_0}(x_0)| < \frac{1}{2}$ implies $|f_{n_0}(y)| < \frac{1}{2}$. But $f_{n_0}(G'_j) = 1$. Therefore $y \in G_j$. That is, for $x_0 \in G_j$, if $||f(y) - f(x_0)|| < \epsilon$, then y is in G_j. $\square$

Bibliography

[Adams, 1994] Adams, C. C. (1994). *The Knot Book: An Elementary Introduction to the Mathematical Theory of Knots*. W.H. Freeman.

[Appel and Haken, 1977] Appel, K. and Haken, W. (1977). The solution of the four-color-map problem. *Scientific American*, 237(4):108–121.

[Bartle and Sherbert, 2011] Bartle, R. and Sherbert, D. (2011). *Introduction to Real Analysis*. John Wiley & Sons.

[Cabrera, 1982] Cabrera, B. (1982). First results from a superconductive detector for moving magnetic monopoles. *Phys. Rev. Lett.*, 48:1378–1381.

[Cohen, 2008] Cohen, P. (2008). *Set Theory and the Continuum Hypothesis*. Dover Publications.

[Conway and Smith, 2003] Conway, J. and Smith, D. (2003). *On Quaternions and Octonions*. Taylor & Francis.

[Crossley, 2006] Crossley, M. D. (2006). *Essential Topology*. Springer.

[Donne and Lush, 1988] Donne, J. and Lush, H. (1988). *No Man Is an Island*. Souvenir Press Limited.

[Doxiadis et al., 2009] Doxiadis, A., Papadimitriou, C., Papadatos, A., and Di Donna, A. (2009). *Logicomix: An Epic Search for Truth*. Bloomsbury.

[Edgar, 2007] Edgar, G. (2007). *Measure, Topology, and Fractal Geometry*. Springer.

[Euler, 1736] Euler, L. (1736). Solutio problematis ad geometriam situs pertinentis. *Comment. Acad. Sci. U. Petrop*, 8:128–40.

[Gamow, 1946] Gamow, G. (1946). Expanding universe and the origin of elements. *Phys. Rev.*, 70:572–573.

[Glashow, 1980] Glashow, S. L. (1980). Towards a unified theory: Threads in a tapestry. *Rev. Mod. Phys.*, 52:539–543.

[Goodman, 2005] Goodman, S. (2005). *Beginning Topology*. Thomson Brooks/Cole.

[Guth, 1981] Guth, A. H. (1981). Inflationary universe: A possible solution to the horizon and flatness problems. *Phys. Rev. D*, 23:347–356.

[Hatcher, 2002] Hatcher, A. (2002). *Algebraic Topology*. Algebraic Topology. Cambridge University Press.

[Herman, 1998] Herman, G. (1998). *Geometry of Digital Spaces*. Birkhäuser.

[Hogben, 1968] Hogben, L. T. (1968). *Mathematics for the Million*. W. W. Norton.

[Krantz, 2009] Krantz, S. (2009). *Essentials of Topology with Applications*. CRC Press.

[Linde, 1982] Linde, A. D. (1982). A new inflationary universe scenario: A possible solution of the horizon, flatness, homogeneity, isotropy and primordial monopole problems. *Physics Letters B*, 108(6):389–393.

[Michaelson, 2013] Michaelson, G. (2013). *An Introduction to Functional Programming Through Lambda Calculus*. Dover Publications.

[Misner et al., 1973] Misner, C. W., Thorne, K. S., and Wheeler, J. A. (1973). *Gravitation.* W. H. Freeman and Company.

[Munkres, 2000] Munkres, J. R. (2000). *Topology.* Prentice Hall.

[Nash, 1997] Nash, C. (1997). Topology and Physics: An Historical Essay. *hep-th/9709135.*

[Penrose, 1989] Penrose, R. (1989). *The Emperor's New Mind: Concerning Computers, Minds, and the Laws of Physics.* Oxford University Press.

[Qadir, 2020] Qadir, A. (2020). *Einstein's General Theory of Relativity.* Cambridge Scholars Publishing.

[Ruffini and Wheeler, 1971] Ruffini, R. and Wheeler, J. A. (1971). Introducing the black hole. *Physics Today,* 24:30–41.

[Salam, 1980] Salam, A. (1980). Gauge unification of fundamental forces. *Rev. Mod. Phys.,* 52:525–538.

[Simmons, 2004] Simmons, G. F. (2004). *Introduction To Topology And Modern Analysis.* Tata McGraw-Hill.

[Steen and Seebach, 1995] Steen, L. and Seebach, J. (1995). *Counterexamples in Topology.* Dover Publications.

[Stewart, 2010] Stewart, I. (2010). *Professor Stewart's Cabinet of Mathematical Curiosities.* Profile Books Limited.

[Ueno, 2003] Ueno, K. (2003). *A Mathematical Gift, I: The Interplay Between Topology, Functions, Geometry, and Algebra.* American Mathematical Society.

[Weinberg, 1980] Weinberg, S. (1980). Conceptual Foundations of the Unified Theory of Weak and Electromagnetic Interactions. *Rev. Mod. Phys.,* 52:515–523.

Index

Accumulation point, *see* Limit point
Adjacent, 7
Al Khwarizmi, 4, 5
Aleph null, 19
Algebra, 4, 5, 11, 104, 177
Algebraic numbers, 21, 46
Arc-wise connected , *see* Path connected
Archimedean property of real numbers, 40

Baire's category theorem, 72, 80, 89
Banach space, 170
Basel problem, 187
bases
    for product topology, 115
Basic open set, 164
Basis, 48, 51, 68
    minimal, 48
    of a vector space, 48
    uniqueness of, 49
    vector, 58
Bijective, 103
Binary operation, 18, 19, 27
Black hole, 179
Bolzano-Weierstrass theorem, 76, 136
Boundary
    as an operation, 46
Boundary of a set, 35
Boundary point, 37, 44
    (exercise), 56
Bounded sequence, 64

Calculus, 47, 58, 109, 130, 142, 150, 171
Cantor set, 89, 163, 169, 177
Cardinal number of a set, 18–20, 22, 34
Cardinality, 21, 24, 25, 35, 46, 59
Cartesian product, 20, 26
Cauchy inequality, 62
Cauchy sequence, 72–76, 78, 79, 84, 86–89, 143
Chaos, 177

Characteristic function, 95
Circle, 100
Clopen, 43, 147
Closed interval, 30, 85, 154
Closed mapping, 101, 113
Closure, 42, 45, 46, 55, 70, 123
    for relative topology, 53
Closure of a set, 36, 37, 44
Closure operation, 103
Co-countable topology, 34, 40, 42
Co-finite topology, 34, 38, 40, 45, 53–55, 134, 153, 165, 166, 169, 170
    basis for, 49
Coarser topology, 50, 51
Comb space, 165, 166
Compactification, 133
Compactness, 131, 132, 135, 136, 140, 142, 144
    locally compact, 144, 145
    sequentially compact, 137, 139, 140, 143
Completely normal space
    (exercise), 128
Component, 160
Composition of functions, 93, 102, 104
Connectedness, 100, 147, 151, 155
    local connectedness, *see* Locally connected
    multiply connected, *see* Multiply connected
    simply connected, *see* Simply connected
Consumer, 9, 58
Continuity, 90, 94
    for functions of several complex variables, 94
    of composite map, *see* Composition of functions
    Union of open sets, 31
Continuous function, 100
Continuously infinite, 24
Continuum hypothesis, 25
Contraction, 82–85
Convergence, 75, 77, 79, 82
Convex (exercise), 169

Corollary, 41
Countability, 51, 105, 106
    first countable, 105
    second countable, 105, 124
Countably infinite, 20, 47
Covering, 126, 131, 132, 134, 135, 146, 164
    sub-covering, 132

Dense, 46
Dense set, 46
Derived set, 44
Diameter, 138
Differential equations, 82, 84, 171
Dimension, 5, 26, 48, 163, 177
    of a vector space, 51
Disconnected, 147
    totally disconnected, *see* Totally disconnected
Discrete metric, 60
Discrete space, 33, 109
Discrete topology, 33, 34, 40, 44, 48, 50, 53, 55, 68–70, 108, 150, 165, 166, 169
Divisor topology, 170
Door space (exercise), 55

Economy
    for five year plan, 61
    of a country, 32
    of the world, 32
Einstein-Rosen bridge, 179
Empty set
    as a closed set, 32
End point, 157
Equivalence class, 86, 87, 163
Equivalence relation, 104, 163
Euclidean topology, 148, 157
Exclusion topology, 33, 132

Finer topology, 50, 51
Finite intersection, 31, 133, 134
First category, 80, 81, 89
First countable, 51, 52
Fixed point, 82
Fixed point theorem, 72
Fort space, 122, 125
    (exercise), 146
Four color map, 7
Fractal, 15, 177
    Mandlebrot set, 178
Frontier of a set, *see* Boundary of a set
Function, 172

Gödel's theorem, 25
Generalization, 108, 121, 131, 147
Generalized Weierstrass theorem, 142
Geometry, 3, 167, 171
Graph, 119
    (exercise), 128
Graph theory, 8
Greatest lower bound, *see* Infimum
Group, 18
Groupoid, 18

Hausdorff, 147
    space, 117, 118, 126, 134–136, 144, 163, 164, 167, 168
Heine-Borel theorem, 131, 136, 143
Homeomorphic, 88, 139, 154, 172
Homeomorphism, 90, 98
Homomorphism, 98
Homotopy theory, 157

Idempotent function, 96
Identity, 92, 104
Identity map, 106
Imbedding, 85, 87
Inclusion topology, 33, 52
Index set, 31, 34, 47, 113, 114
Indiscrete space, 33
Indiscrete topology, 33, 40, 46, 55, 108, 113, 165
Induced topology, 52
Infimum, 64, 117, 123, 130, 131, 142
Inherited topology, *see* Induced topology
Initial point, 157
Initial value problem, 85
Injective, *see* One to one, 98
Inside, 10
Integers, 151
Interior, 15, 35, 36, 41, 42, 55, 64, 71, 93
    of a set, 35, 37, 42
    point, 35
Intermediate value theorem, 147, 150
Into, 102
Invariance, 10
Inverse, 18, 19, 91, 99, 136
Irrational numbers, 5, 23
Isolated point, 40, 89
    for a discrete space, 44
    for an indiscrete space, 44
    of a perfect set, 44
Isometry, 85, 86
Isomorphism, 98

Jordan, 147

Königsberg bridge problem, *see* Seven bridges problem
Klein bottle, 13
Knastern, 147
Koch curve, 15
Kronecker delta, 99
Kuratowski, 147

ℓ_2, 76
ℓ_p, 77
ℓ_∞, 78
Larger topology, *see* Finer topology
Largest open set, 35
Law of the excluded middle, 25
Least upper bound, *see* Supremum
Lebesgue number, 140
Left ray topology, 34, 70, 111
Lemma, 41
Limit point, 43–45, 53, 55, 71, 74, 75, 113, 136, 137, 140
Lipschitz condition, 85
Lobachevskian geometry, 46, 60
Local bases, 122, 124
Local basis, 51, 52
Locally compact, 145
Locally connected, 165–167
Logarithm, 4
Lorenz attractor, 178

Möbius strip, 13, 176
Manifold, 139
Map, 8, 93, 106, 118
Mathematics
    application in Science, 3
    as a language, 2
Maximal connected subspace, 160
Metric, 58, 64, 72, 85, 136
    square metric, 61, 63, 65, 70, 96
Metric space, 60
    complete, 73
Metrizability, 106, 124
Metrizable space, 68
Minimal, 140, 170
Minkowski geometry, 60
Minkowski inequality, 63
Monarchic topology, 109
Monoid, 18
Monotonic operations, 45

Morphism, 90
Multiply connected, 26, 173

Necessary condition, 41
Necessity, *see* Necesssary condition
Negation, 17, 18, 25
Neighborhood, 37, 38, 165, 166
    of a set, 47
    notation, 37
        for relative topology, 53
Neighborhood system, 38
Norm, 62
Normal space, 124, 126
    Completely normal space, *see* Completely normal space
Normed space, 30
Nowhere dense, 79
    (exercise), 56
Nowhere differentiable, 81

Oligarchical topology, 128
One to one, 18–22, 24, 98
Onto, 25, 26, 98, 102, 149, 150, 156, 159, 167, 168
Open cover, *see* Covering
Open interval, 30–32, 35, 40, 65, 66, 70, 89, 107, 121, 128, 134, 165
Open set, 30, 66
Open sphere, 64
Order relation, 104
Ordering, 11, 12, 22, 24, 26, 27
Orthogonal transformation, 99
Outside, 10

p−exclusion topology, 125
p−inclusion topology, 152
Paradox, 16, 17, 32
    Russell's paradox, 16
Partial order, 27
Pasting, 14, 96, 97
Patching, 97
Path connected, 106, 157–160, 167, 169, 170
    in $\mathbb{R}^n \smallsetminus \{O\}$, 158
    Topologist's sine curve, 159
Perfect set, 44
    (exercise), 55
Planar curves, 39
Position vector, 58
Positive definiteness, 60
Power set, 24, 32, 48
Producer, 9

Product topology, 114
Projection mapping, 96, 101, 119
Pseudo–metric, 60

Rational numbers, 149
Reductio ad absurdum, 35, 94, 137
Reflection transformation, 99
Reflexive relation, 104
Regular space, 121–124, 127
Regularity, 123
Relative topology, *see* Induced topology
Relativity theory, 60, 178
Restriction, 104, 116
Riemannian geometry, 46, 58, 139, 143, 145
Right ray topology, 34, 70, 111, 113
Rotation transformation, 99

Second category, 80, 81
Second countable, 51, 52
Self-similarity, 178
Semigroup, 18
Separable, 47, 48, 51, 105, 109, 120, 127, 167
Separated, 154
Separation, 147
Sequence, 46, 156
    convergent, 73
Set
    definition of, 15
Seven bridges problem, 6
Simply connected, 26, 173, 175
Smaller topology, *see* Coarser topology
Smallest closed set, 35, 45
Songfrey line, 52
Sphere, 57, 100
Stereographic projection, 25–27, 100, 172
Subspace, 52
Sufficiency, *see* Sufficient condition
Sufficient condition, 41
Superset, 45
Supremum, 64, 130, 138, 142, 154
Surface, 5, 6, 8, 14, 57, 58, 100, 173, 175
    of a sphere, 175
Surjective, *see* Onto
Symmetric relation, 104
Symmetry property, 60

T_0 – space, 109, 110, 112, 113, 120, 121, 129
T_1 – space, 109–114, 121, 122, 125, 126, 128, 129
T_2 – space, 109, 113, 116, 118–120, 122, 128, 146
T_3 – space, 120, 121, 124

T_4 – space, 121, 125
Theorem, 41
Topolgical invariant, *see* Topological property
Topological embedding, 104
Topological property, 105, 113, 124, 135, 150, 159
Topological space, 32
Topologically equivalent, 172, 173
Topologist's sine curve, 158, 159
Torus, 13, 14, 26, 148, 167, 173, 176
Totally bounded, 138–143
Totally disconnected, 163, 164, 169
Transfinite numbers, 24, 25
Transitive relation, 104
Translation, 96, 99
Triangular inequality, 63, 71, 140
Triangular property, 60

Uniform continuity, 142
Urysohn, 147
    Urysohn's lemma, 187
        (exercise), 128
    Urysohn's metrization theorem, 124
Usual metric, 60, 62
Usual topology, 32

Well ordered, 11, 24
Winding number, 176
Wormhole, 180

Don't be Deceived

Book Piracy and Plagiarism are Crimes.

Buy Genuine Oxford Books

Look out for the new security label whenever you purchase an Oxford textbook or supplementary reader. Labels with the features shown below are proof of genuine Oxford books.

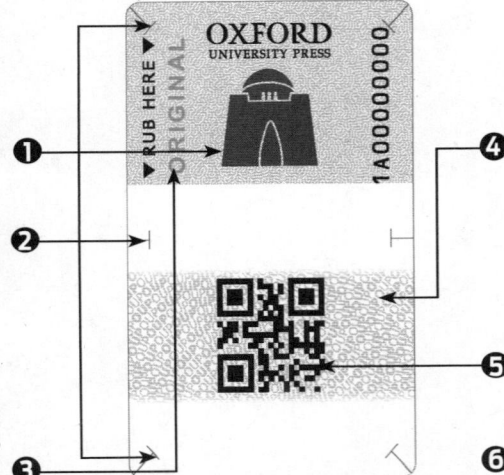

The image of Quaid's Mausoleum changes colour from orange to green when viewed from different angles. ❶

The labels have security cut marks to prevent them from being peeled off and reused. ❷

The word 'ORIGINAL' appears when the area under 'RUB HERE' is rubbed with a coin. ❸

'OUP' written in very small print becomes visible when viewed through a magnifying glass. ❹

- Download free 'FPQR Tasdiq' app from App Store/Play Store on your smartphone.
- Scan QR code on the security label to verify if you have purchased a genuine Oxford book. ❺

The labels tear if peeled off from the book cover. ❻

Pirated books can be recognised by:
- inferior production quality
- low-grade paper
- variations in texture and colour
- poor print quality
- blurred text and images
- poor binding and trimming
- substandard appearance of the book

Do not accept the book if the label is missing, has been torn or tampered with, the colour on the security label does not change, or the word 'ORIGINAL' does not appear upon rubbing the area with a coin.

If you suspect that you are being sold a pirated book without the security label, please contact:

Oxford University Press,
No. 38, Sector 15, Korangi Industrial Area, Karachi-74900, Pakistan.
Tel.: (92-21) 35071580-86 Fax: (92-21) 35055071-72
Toll-free No.: 0800-68775 (9 a.m. to 5 p.m.; Monday to Friday)
Email: central.marketing.pk@oup.com
Website: www.oup.com.pk Find us on: